The Hans Legacy

The Hans Legacy

A Story of Science

Dodge Fernald
Harvard University

Illustrated by James Edwards

LAWRENCE ERLBAUM ASSOCIATES, PUBLISHERS
1984   Hillsdale, New Jersey   London

Lawrence Erlbaum Associates, Inc., Publishers
365 Broadway
Hillsdale, New Jersey 07642

Library of Congress Cataloging in Publication Data

Fernald, L. Dodge (Lloyd Dodge), 1929-
The Hans legacy.

Bibliography: p.
Includes index.
1. Psychology—History. 2. Psychology—Methodology.
3. Clever Hans (Horse) 4. Freud, Sigmund, 1856-1939.
Analyse der Phobie eines fünfjährigen Knaben.
5. Behaviorism (Psychology) 6. Psychoanalysis.
I. Title.
BF105.F47 1984   150'.9   83-11539
ISBN 0-89859-301-8
ISBN 0-89859-413-8

Printed in the United States of America
10 9 8 7 6 5 4 3 2

Preface

An old adage guided the writing of this book: to convince people, tell them a story or give them the facts. This book does both.

The story is that of two characters called Hans and the facts are those of psychology. The aim is to show the reader that the study of science can be a useful, engaging experience.

So go gently. Enjoy this true story. The facts should speak for themselves.

D.F.
Cambridge, Massachusetts

Contents

The Hans Legacy

City Streetlight

Prologue

Once upon a time there were a horse and boy named Hans. The horse was unusually sensitive to human beings and, by coincidence, the boy was unusually sensitive to horses.

Their tales from two cities, Berlin and Vienna, follow the same path. Together, they tell a single story—the story of science.

Griebenow Courtyard

Part One
An Early Debt

CHAPTER ONE

Origins

North of the heart of the city, down on Griebenow Street, there was a small, open courtyard. To this rather forlorn place some years ago a man and his horse came daily to toil at a most unusual task. Their work seemed hopeless, so much so that spectators gathered on the balconies and at the windows of the nearby tenements to jeer and make catcalls during the proceedings.

The man, approaching his seventies, had never married and found bachelorhood well suited to his present purpose. With no family and few friends, he spent most of his time working with his beloved, generally docile animal. He cut quite a figure in that Berlin courtyard, with his long white duster and floppy white hair sticking out beneath a broad-brimmed, dark hat, which he wore even on the warmest days.

The horse was a large brown stallion of Russian breeding with white socks and a stately bearing. Its proud carriage and intelligent eyes made it seem especially suited to carry a difficult burden, not on its back but in its head. This horse, by the most modern methods available, was being carefully schooled in the rudiments of human intelligence. His owner, a teacher of mathematics in earlier days, was training the animal in music, reading, spelling, and numbers, as well as use of the clock, calendar, and coins.

Working usually at midday in the heat of the sun, brisk winds, or other impediments of the open area, the old schoolmaster at first found no reward. Progress was intolerably slow, despite high hopes and his obvious teaching skills, and after two years of effort he advertised the horse for sale. But soon his interest revived; he returned to the horse; and his work began to prosper. There was an iron will in that thin, bent body, and step by step he nurtured the animal's natural abilities. Someday, through his patience and powers, he would make believers of those who had come only to ridicule.

And he did, at least in part. After another couple of years his horse became front-page news, solving not only typical school problems but questions to which even scholars gave erroneous answers. For many honorable persons, including those who considered themselves specialists, his achievement marked the beginning of a new era in our understanding of human and animal behavior. The borders among the species would fade. The phenomenal horse would change forever our view of the limits of animal intelligence.

But the question remained. How did he do it? What had made this horse called Hans so extraordinary?

We shall return shortly to the feats of this learned quadruped, and to the commotion they caused, but at this same moment in history, April 1903, just when the old man began working with his Hans in Berlin, a father and mother in Vienna commenced a similar task. Down on Lower Viaduct Street they began training their own Hans, a newborn son. Unlike the aging schoolmaster, they were not planning anything spectacular; they simply wanted their first child to develop into a normal, healthy adult, avoiding insofar as possible the usual conflicts and fears of childhood.

The father, a bespectacled, mustached physician, was particularly interested in problems of adjustment; the mother, years earlier, had suffered a mild form of maladjustment for which she and her husband had successfully sought help. Thus, they decided to take special pains with this child, making careful observations and keeping records of his behavior. They would raise him with little coercion, using no more restraint than absolutely necessary. Should problems arise, they would be handled by reference to the records and, if necessary, outside assistance.

As the boy matured, he became a precocious lad with a lively curiosity, exploring everywhere and constantly asking questions. He inquired about the origin of children and was told the story of the stork; he asked about his parents' rules and was encouraged to be independent.

One day he announced that he was going to stay with a friend. "Well," responded his mother, "if you really want to go away from Daddy and Mummy, then take your coat and knickers and—good-bye!" The boy gathered up some clothes and started for the stairway, at which point he was quickly retrieved.

He played with children his own age and jokingly referred to two friends as his own offspring. "My children Berta and Olga were brought by the stork too," he announced. He had imaginary play-mates, feared falling into the bathwater, and at times became so angry that he wanted "to spit." But altogether he seemed to be a normally developing child. A friend of the parents described him as a bright and happy little boy.

It came as a great surprise when one day this calm, self-assured little chap suddenly found himself highly agitated. Soon he became unable to take naps or to go to sleep readily at night. And eventually, as his condition worsened, he even became upset at the thought of leaving the house.

What had happened? Why was the child so afraid, despite the most conscientious efforts of his parents to prevent this outcome?

This, then, is the Hans legacy, the story of a horse and boy named Hans. The horse possessed an unusual intelligence, astounding experts of his day; the boy possessed an unusual personality, leaving those who knew him highly perplexed. Both caused considerable controversy, some of which remains today, and the study of both individuals contributed to our understanding of the research process. From these investigations there came a legacy, something of value to be passed to the next generation, and it has proven significant in all of science, not just the field of psychology.

Founding of Modern psychology, the framework for this story, began a gener-
Psychology ation earlier, in late nineteenth-century Europe, by which time
Charles Darwin's ideas had gained considerable acceptance. All
highly developed organisms, according to his theory of evolution,
developed from lower forms of life. Each species, including human
beings, evolved from a more primitive animal ancestry and has
done so according to the principle of the survival of the fittest
individuals.

This approach to the origins of the species prompted increased
interest in animal behavior. People asked: If we have minds, and
animals are our ancestors, do animals have minds like ours? If
animals have instincts, do we have instincts too? Darwinian biology was a most important precursor of modern psychology in North
America and Europe, and certainly it was the most significant
factor in the rise of animal psychology.

In this same way it contributed to the Hans legacy. The eccen-
tric schoolmaster was inspired to undertake his tedious instruction
of the horse chiefly by Darwin's work. If the horse and humanity
both have an animal ancestry, he decided, they must have a comparable intelligence. And his remarkable result rekindled Darwinian debates all over the Western world.

Another contributor to the development of modern psychology
was experimental physics, which provided methods for measuring
various forms of stimulation, including light, sound, and touch.
These investigations showed that precise changes in stimulation
prompted precise but not necessarily predictable changes in a person's response. For example, when identical silver coins were
placed on a prone person's forehead, one cold coin often was perceived as heavier than two warm ones, one atop the other, and this
type of sensitivity varied widely in different regions of the skin.

Once adopted from physics, these experimental methods proved
extremely useful in the emerging field of psychology. They constituted, in fact, the chief procedure for studying the intelligent
horse. In turn, developments in theoretical physics influenced the
investigator who studied the fearful boy.

In varying degrees, biology, physics, and also physiology contributed to the founding of psychology. A few years earlier, physiolo-
gists had developed a method for measuring the speed of the nerve

10

impulse, thereby demonstrating an irreducible interval between the stimulation and response. A frog's leg was stimulated close to the muscle and then further away, and the difference in time until the muscle contraction in each case was the time required for neural transmission between those points. Such studies revealed important relationships between physical structure and behavior, and they also indicated that controlled procedures and even mathematical principles could be usefully applied in psychology.

Eventually, a German physiologist and philosopher, Wilhelm Wundt, was credited with the founding of psychology. At the University of Leipzig he developed a method for conducting experiments on seeing, hearing, and other sensations, and in 1879 his first graduate student completed a research project. Wundt thereby decided that he had established a psychological laboratory, and with this emphasis on research, psychology came into being.

Wundts Upon a Birthday. *Husband and wife, in the middle of the second row and surrounded by psychologists, celebrate Wilhelm's eightieth year.*

The field of *psychology* is defined as the scientific study of human
and animal behavior and experience. Psychologists investigate the
responses of living organisms at all phylogenetic levels, and they do
so by scientific procedures. One of our most recent and most important enterprises in science is this study of our own actions and
feelings.

Considered in this way, psychology is a *basic science*, devoted to
the acquisition of new knowledge. The aim is to uncover fundamental principles of behavior and experience. Knowledge is sought
for its own sake, not for its usefulness. But psychology is also concerned with practical issues and regarded as an *applied science*. The
aim here is to use psychological knowledge in the solution of daily
problems, including education, business, health, law, and even the
use of leisure time. Psychology is most visible in its applied forms,
but its success in this area inevitably depends upon advances in
basic research.

The question of the horse involved psychology as a basic science.
It was simply a problem that people wanted solved, like a mountain
to be climbed, and basic research methods were used for this purpose. The question of the boy illustrates applied psychology. It was
a practical matter involving someone's health and happiness, and
techniques of applied psychology were used here. Together, these
questions, basic and practical, constituted an early challenge for
the emerging field, scarcely a generation after its founding. They
would test its capacity to solve problems of broad public interest.

The horse called Hans, proclaimed by some observers as the
miracle of the century, could not be dismissed with a disrespectful
shrug of the shoulders. Something had happened in the old man's
work, and the ensuing debate raged for months, matched in intensity only in heated political campaigns. In earlier ages the animal's
industrious, cantankerous owner surely would have been burned as
a witch, for the suspicion created by his work could hardly be
imagined.

The boy called Hans had been a "perfectly reasonable little
member of human society," reared in comfortable circumstances by
an apparently loving family. Who would have predicted a difficult
time for this happy, precocious child? Who would have guessed

that soon he would be compulsively avoiding something that frightened him? His behavior befuddled his parents and their friends; for a time it remained a mystery even to the boy himself; and eventually some experts decided that it was an inevitable consequence of all family life in Western society.

Emerging psychology, confronted with these problems, was prompted and prepared to accept this challenge partly by the aforementioned developments in other fields—Darwinian biology, experimental physics, and nineteenth-century physiology. They gave psychology some early objectives concerning human and animal behavior and also certain methods, including new experimental techniques. These fields, then, are bound into psychology and the Hans legacy, directly and indirectly, in those subtle, inexplicable ways that tie all our lives together. There were other precursors as well, for history is inevitably without clear beginnings, but these are the fields to which psychology owes its earliest debt.

Near Unter den Linden

Part Two
Hans in Berlin

CHAPTER TWO

A Question

Berlin at the beginning of this century was growing at a remarkable rate. Kaiser Wilhelm, the youthful German emperor, and the Industrial Revolution, underway for more than a century, had been leading the city to unprecedented industrial and commercial prominence in Europe.

The population had increased tenfold in less than a century, and it would double again in the next generation. There was prosperity in the iron and steel works, the cloth trade, and the well-known breweries. Also the shopkeepers thrived on the demands created by the new laborers. Pedestrians and horsedrawn vehicles filled the streets, adding to the din of noisy parades for the ambitious Kaiser.

These colorful displays of German strength and unity came marching down Unter den Linden, a broad boulevard in the center of the city named for its place under the linden trees. Brass bands and marshals on prancing horses led armies of soldiers and citizens to celebrations in the Lustgarten, a large square at the eastern end of this spacious avenue. In those days of high employment, low prices, and confidence in the future, such demonstrations were expected.

At the center of this vibrant enterprise was an efficient transportation system. It included the new Berlin subway, opened in 1902, an intricate canal network, and modern highways. On these roads soon would be seen all manner of early automobiles, for the gasoline engine had advanced to four cylinders and pneumatic tires had become available. Through the perseverence of its citizens, the

Changing Times. *Early motorists, stranded by their automobiles, listened to the derisive shout of buggy drivers.*

city could even lay some claim to the development of the automobile industry.

This rapid economic growth also induced cultural advances. Oratory and drama were prominent; new libraries had been constructed; and newspapers had a wide circulation. Even the views of the controversial Englishman, Charles Darwin, were more readily accepted here than abroad. The first translation of his work appeared in German, and a national literary movement called naturalism adopted the basic Darwinian principles. German professors were widely acclaimed, especially in philosophy, physiology, and the struggling new enterprise of psychology.

At the University of Berlin, and also among the pedestrians on Unter den Linden, there was by 1904 considerable talk of national and international issues: the Kaiser, German colonization, the miners in the Ruhr valley, and that extraordinary animal down on Griebenow Street. Almost overnight this trotting horse had become a sensation, not for races won at the track but for problems solved in a decaying courtyard on the northern edge of the city.

This perplexing creature, appropriately called Clever Hans, could spell and read cursive German script; he could do mathematics; he could discriminate colors; he could tell time; and he could give demonstrations of a marvelous memory. He even exhibited musical skills and identified numerous objects.

"How many of the gentlemen present are wearing straw hats?" the horse was asked.

Clever Hans tapped the answer with his right foot, being careful to omit the straw hats worn by the ladies.

"What is the lady holding in her hand?"

The horse tapped out "Schirm," meaning parasol, indicating each of the letters by means of a special chart. He was invariably successful at distinguishing between canes and parasols and also between straw and felt hats.

More important, Hans could think for himself. When asked a completely novel question, such as how many corners in a circle, he shook his head from side to side to say there were none.

For anyone in town it certainly was worth the effort to see this animal perform, since no fee was required. Almost any spring or summer day in 1904 one merely had to stroll into the spacious courtyard on Griebenow Street and take a seat in the small gallery or stand aside with other spectators, among whom one might find famous scholars and foreign visitors. Usually Hans stood quite still, about six meters from the audience and directly in front of his work tables, his long mane and tail gently blowing in the breeze. His elderly trainer stood just to his right and a bit in front of him.

When all was quiet, the performance began. Hans was asked questions first by his master and then by members of the audience. In one instance it was decided that Dr. Grabow, a retired member of the Berlin schoolboard, should sing to the horse, as a test of its musical ability. The good doctor sang two notes and then the horse was asked: "How many intervals lie between?" Clever Hans, possessing perfect tonal discrimination, tapped twice, which was the correct answer. The animal had an unerring capacity for knowing the proper intervals in chords, making consonance out of dissonance.

When the tones C, D, and E were presented simultaneously, Clever Hans was asked: "Does that sound pleasant?" He shook his head from side to side. "What tone must be omitted to make it pleasant?" The horse tapped out "D." With a more complex chord, D-F-A-C, the horse indicated that the C should be eliminated, showing a musical inclination that did not favor minor seventh chords. Altogether, Hans was familiar with thirteen different melodies.

But nothing more effectively established the horse's reputation than those well-documented instances in which Hans was correct, rather than his human questioner. One day, on what he thought was the seventh of the month, a man entered the horse's stall and asked the date. Clever Hans tapped eight times and suddenly Count Otto zu Castell-Rüdenhausen, a prominent figure in Berlin, realized to his embarrassment that it was indeed the eighth day of September.

On another occasion this same gentleman asked the sum of 5, 8, and 3, expecting the answer 10 because he thought the second digit had been a 2. To his surprise and later chagrin the horse tapped 16.

A local newspaper reported how Clever Hans was requested to

20

spell "Dönhoff." Using the special chart prepared by his master, he began with "Dö." His questioner interrupted, calling for "o" instead of "ö," but the horse continued, spelling the full name correctly. Hans had not erred, but rather his questioner, who was thinking instead of the name "Dohna."

Such consistent and notable success in many areas of human knowledge impressed the curious and credulous, who came in increasing numbers to observe the amazing stallion on Griebenow Street. Clever Hans was a case of special intelligence. But what is this thing called intelligence?

This question has provoked debate for centuries, but one definition that was current at the time of Clever Hans still survives today. It states that *intelligence* is the capacity to acquire and use information—a definition that has the advantages of being broad and readily understood. Intelligent people are those with the greatest amount of knowledge and greatest ability in using it. They have learned about the world and can reason with what they have learned.

But there is also a limitation. Human beings live in different environments and know different things. No one can know everything and therefore a decision arises concerning which knowledge is most important. Mathematical reasoning is emphasized in urban classrooms, and it is essential in business. Knowledge of the soil, wildlife, and weather is indispensable in rural life; without it, and the ability to reason, no one could survive for long.

This problem of deciding which knowledge best reflects intelligence has led to another definition: intelligence is the capacity to adapt to the environment. This definition recognizes that human beings live in diverse environments and therefore possess diverse kinds of knowledge, and it is applicable across the phylogenetic scale. The worm is intelligent in the mud, the fox in the woods, and the steeplejack on tall buildings. Human beings, acknowledged to be the most intelligent species, live in the widest variety of environments, ranging from the bottom of the sea to the highest mountains and driest deserts. But there is a problem here too. How

21

do we compare individuals from different environments? Which criteria should be used? Almost any decision favors one culture or another.

By either definition Clever Hans appeared intelligent. He displayed a broad range of information, and he had adapted to the normal horse environment, as well as to aspects of the human world.

The horse, for example, could use a watch, and he could even answer questions about the face of a watch. "Between what figures is the small hand of a watch at five minutes after half-past seven?" "How many minutes has the large hand to travel between seven minutes after a quarter past the hour and three-quarters past?" Hans sometimes answered without looking at a watch, which demonstrated memory and reasoning, not merely counting.

Problems in definition, it should be added, occur with all our important concepts, including success, happiness, justice, and even life itself. This circumstance does not prevent us from using these ideas, however, or from studying the phenomena to which they refer. In fact, while Clever Hans was tapping in that Berlin courtyard a French psychologist, Alfred Binet, was making notable progress in the measurement of intelligence. He had been asked to identify schoolchildren who had insufficient ability to profit from normal classroom instruction, and for this purpose he developed the first practical intelligence test.

Components of Intelligence   In preparing this instrument, Binet began with the premise that many different abilities should be involved, such as perception, memory, use of words, and numerical reasoning. He then developed a wide variety of test questions for the different age levels. These questions contained many defects, but they were revised and refined until the original aim was reasonably achieved. The result, in 1905, was called the Measuring Scale of Intelligence, and it helped shape our modern conception of intelligence as a multifaceted capacity.

Animals, young children, and retarded adults, at the lowest level of ability, are tested with problems involving *simple discrimination*, in which the subject chooses among two or more simple alternatives. In a test of musical discrimination, Clever Hans was presented with one tone and then another immediately afterward. The horse was to indicate whether the second tone, such as middle C, was the same as the first.

In testing for color discrimination, cloths of various hues were randomly arranged on the ground and Clever Hans was asked, for example, to select the red one. He would pick it up in his mouth and present it to his questioner. Sometimes the cloths were dangled from a string tied between two poles, but the procedure made no difference—the horse succeeded anyway.

A cavalry officer once stood before the horse and asked Hans the color of his cap. The clever animal stamped his foot three times, indicating that the cap was red, like the third cloth.

Somewhat more difficult are *memory tasks,* in which the subject is required to observe something and to remember it for varying intervals. The stallion was shown a photograph and shortly thereafter five members of the audience assembled before him. He was asked to select from among these people the one whose photograph he had seen previously, and he readily gestured toward the correct person with a movement of his head. When told the values of German coins, he remembered them all later, and he rarely seemed to forget a person's name. He was introduced to a reporter from the *New York Times,* Edward Heyn, and when the man appeared again in the courtyard, Hans spelled Mr. Heyn's name before it was mentioned. He even recognized in civilian clothes a man previously introduced in uniform.

We should note that Han's mode of responding was somewhat the reverse of ours. He pointed with his mouth and talked by tapping his hoof. He indicated monetary values, for example, by one tap for gold, two for silver, and three for nickel. With playing cards, the system was one tap for hearts, two for clubs, and so forth, and when words were necessary, the special spelling chart was used.

Some spectators, after witnessing these remarkable feats, believed that the horse was a divine creature, but they were quickly dissuaded when Hans reared on his hind legs, snorted, and pawed in the air at a great height, seemingly angry at his questioner. His master explained that the horse, when he refused to answer, had an unruly temper. Fortunately, these instances of misbehavior were infrequent.

There was an obviously affectionate relationship between master and beast. From time to time Clever Hans nuzzled his trainer, apparently as a friendly reminder to keep things moving, especially when the man had something tasty in his pocket. The horse once laid his thick lips on the back of another man's neck, giving the suggestion of a kiss, and after this demonstration of tenderness it did not seem appropriate to withhold the carrot.

But tasks of memory and simple discrimination were not all the horse could do, as we have seen. Like older children and adults, he also showed *language ability*, using various words. When asked where he lived, Hans spelled "Krippe," meaning manger. He was shown a picture of himself and asked, "What is this?" The animal tapped out "Pferd," meaning horse.

It was astounding how Hans could repeat perfectly a sentence told to him one day previously. Once he was told: "Brücke und Weg sind vom Feinde besetzt." The next day, using his special chart, he made the necessary taps to spell this sentence correctly, which in English means, "The bridge and the road are held by the enemy." The horse performed these tasks for his indefatigable trainer, Mr. Wilhelm von Osten, and also for dozens of different questioners, including people from the audience.

Eventually, Mr. C. G. Schillings, the well-known African explorer, came into the courtyard, arriving in an apparently skeptical frame of mind. He asked permission to work alone with Clever Hans, and he too achieved remarkable results. The horse was phenomenal, he declared, and afterward he spent much time displaying Hans to guests. A zoologist by training, Schillings was so convinced of Clever Hans' abilities that he sent a special report to the Sixth International Congress of Zoologists, meeting in August in Switzerland. This account was most enthusiastic and suggested further inquiry.

Schillings' reputation brought still more fame to Clever Hans, and soon privy counselors, ministers, army officers, and athletes were numbered among the horse's advocates. The case of the wonderful horse was widely acclaimed.

On occasion Hans did make an error, even under Schillings' tutelage. Sometimes the questions were too numerous and the horse became nervous or inattentive. Sometimes he refused to answer, as happened with an officer wearing a monocle and twisted mustache, which seemed to offend Hans.

"Now pay attention!" Mr. von Osten commanded in such instances. Then the question was presented again.

The horse often had difficulty with the number one, tapping twice instead. This rather common error in the simplest of problems baffled his owner and further convinced him that the animal had a will of his own.

Crowds in the Courtyard. *As policemen kept order, Mr. Schillings displayed Clever Hans for an elegant audience from the west side of Berlin.*

Sometimes Clever Hans showed his stubbornness by repeating a wrong response or persisting in a formerly correct one even after a new question had been asked. Once when Mr. Schillings and Mr. von Osten asked the same question before a large gathering, the horse answered Mr. von Osten correctly, tapping twice, but persisted in giving three taps to Mr. Schillings. Suffering this indignity

25

three times in a row, the ruffled Schillings exclaimed: "And now are you going to answer correctly?" To the merriment of those in the grandstand, Hans shook his head.

Indeed, the spectators often became involved in the horse's performance. When Mr. von Osten held up a slate with the number 13, the horse once answered with five taps. "Wrong!" shouted those in the grandstand. But Clever Hans usually corrected himself when questioned again. These visitors, taking advantage of some momentary absence on the part of his trainer, sometimes put a hasty question to the beast, which Hans answered correctly with amazing regularity.

So great was the horse's reputation that no one was surprised when the Prussian minister paid his respects to the learned mammal. "Why shouldn't he?" wrote one wag. "A horse cannot ask for a position!"

The Duke of Sachse-Coburg-Gotha came to see a performance, and Hans gave even that man's name correctly, surpassing most members of the audience. On that same day Mr. Edward Heyn once again visited the Griebenow courtyard, and in the *New York Times* Sunday issue of August 14 he wrote: "By the time this article is printed the Kaiser, who has heard with interest of this horse prodigy, will have seen the animal." The Kaiser never came in public, but Clever Hans continued to prosper.

Of all his abilities, Hans' most remarkable talent lay in *mathematical reasoning*, which is thinking with numbers. Certainly he could count and distinguish among items. When several girls and officers stood in a line, he indicated the number of each. But he was equally comfortable with basic arithmetic, fractions, and decimals, even in the abstract. "How much is ⅖ and ½?" the horse was asked. He tapped out nine and then ten, meaning ⁹/₁₀. With fractions he always tapped the numerator first, then the denominator.

Another problem was posed: "What are the factors of twenty-eight?" Hans tapped 1, 2, 4, 7, 14, and 28, which is correct.

The horse also answered a variety of calendar questions: "If the eighth day of a month comes on Tuesday, what is the date of the following Friday?" He indicated the correct day of the week for anyone's birthday, and it seemed he did the problem mathematically, calculating the number of days in each month. Or else he remembered the whole calendar for a given year. In either case, his success was astounding.

This range of problems does not identify all those administered to the horse or even those in modern tests of intelligence. But it does illustrate the horse's ability and emphasizes that two general types of items are found on the most well-developed intelligence tests. Some items, called *performance items*, do not require language ability. The subject simply answers by pointing, drawing, assembling something, and so forth, as the horse did in selecting a cloth of a certain color. These items are particularly useful in assessing the mental ability of foreigners, poorly educated persons, and young children. In addition, there are *verbal items*, requiring the subject to write or speak the answer. Here Hans was given a concession; he merely tapped out "Pferd" and other words. Language is a most important tool for adaptation and problem solving in the human environment.

While using these kinds of questions, Alfred Binet suggested a useful convention for presenting the results. He described his findings in terms of *mental age*, which is the normal or usual level of mental development for a given chronological age. A mental age of seven, for example, meant that the child had passed the same test items as the average seven-year-old. A child with a chronological age of seven and a mental age of 14 would be a bright child, having reached the mental development of the average 14-year-old. But with the mental age of four, this same child would be considered retarded. The concept of mental age provides a means by which individuals of different chronological ages can be compared with respect to normal intelligence.

Eventually, Binet's method was adapted throughout the world, and it provided the basis for a more facetious definition: intelligence is whatever intelligence tests measure. This test was not readily available in Clever Hans' time, however, and educated persons simply estimated the horse's mental age. It was agreed that his mind was equivalent to that of a 13- or 14-year-old person, at least.

With this level of intelligence, the horse became a celebrity in all parts of the Western world. There were books, pictures, and magazine stories. Advertisers used his name for selling toys and drinks, and people wrote songs and plays about Clever Hans. In late August, letters to the editor of *Berliner Tageblatt*, a leading city journal, reached "downright alarming proportions." The miracle of Clever Hans shared headlines with domestic and foreign issues, including nationalization of the German mines and the Russo-Japanese War.

Even the stable boy was quotable. Perhaps seeking publicity, he told the newspapers that Clever Hans was not responding to Mr. von Osten's words. Rather, the horse was responding to the boy's eyeblinks. The secret lay in his eyes! And with this disclosure academic and lay persons began taking sides.

Support for the horse was especially prevalent among those who had seen a performance. These people usually claimed that Clever Hans was capable of all the feats attributed to him, and most of Mr. von Osten's neighbors on Griebenow and Wolliner Streets were in this category. Besides the redoubtable Mr. Schillings, another prominent advocate was General Zobel, who had entered the controversy a whole year earlier.

At that time, Mr. von Osten wanted to demonstrate the worth of his steed and he placed a newspaper invitation to observe his work. Interested parties were to arrive at 10 Griebenow Street at 11 o'clock in the morning, and General Zobel appeared on the scene. He tested the horse in the absence of its owner, obtained successful responses, and wrote a detailed report. It was spurned by several editors, but in the following July the general finally published his piece and Clever Hans gained further credibility.

A most important article on Clever Hans appeared in Italy, written by a man named Emilio Redich. It prompted a painting of the horse, more stories, and a large following among the Italians.

But other people called the horse a fake, a clever circus act, a pack of lies. A cavalry officer wrote from Russia that the whole venture was impossible. Another man said that the animal did not have the sense of a three-year-old. Still another doubted the horse's abilities but praised the careful methods by which Mr. von Osten had instructed the beast.

Nr. 439. XXXIII. Jahrgang

Abend-Ausgabe.

Montag, 29. August 1904.

Berliner Tageblatt
und Handels-Zeitung.

Die Notstandstariffrage.

Offiziöse Federn setzen ihre Kampagne zur Verteidigung der preußischen Eisenbahnverwaltung in der Notstands-tariffrage mit einer Zuversichtlichkeit fort, die er-kennen läßt, daß man in Regierungskreisen die Argu-mente der Fürsprecher der Bahnbeförderung und Industrie so recht gern durch die Zahl zu erreichen genau weiß, was die bereits die Zeit der offiziösen Argumentation Selbstverständlich bringt jeder Selbstverständlich bringt jeder wieder mit derselben Gegenargumente tarife vor, die seine Vorgänger seit unverzählt haben...

Das Pferd des Herrn v. Osten.

Zwei Urteile. [Nachdruck verboten.]

Das Interesse für den „klugen Hans" in der Griebenow-straße nimmt, wie zahlreiche Zuschriften aus unserem Leser-kreise beweisen, geradezu beängstigende Dimensionen an. Es wird Zeit, daß man endlich einmal fernstehende und kalt urteilende Personen mit klaren Worten aussprechen, was eigentlich an dem „Wunder" ist. Das kann aber natürlich nur einen Zweck haben, wenn die Beurteiler das Pferd wiederholt ge-sehen und beobachtet haben; denn mit dem gering-schätzigen Achselzucken von Sportsmännern und Kunst-reitern, die den „klugen Hans" nicht kennen, ist es in diesem Falle nicht getan: diesen Fachleuten stehen doch zu viele ehrenwerte Männer gegenüber, die sich auch Fachleute nennen dürfen, und die auf Grund eigener Anschauung eine ganz andere Ansicht von der Sache gewonnen haben. Von dem Widerstreit der Ansichten geben zwei kleine Aufsätze Zeugnis, die wir hier veröffentlichen wollen. Der Verfasser des einen, ein junger Berliner Gelehrter, ist völlig überzeugt und sieht die Taten des „klugen Hans" bereits für den Beginn einer neuen Aera wissenschaftlicher Forschung an. Der andere Beobachter ist Professor Dr. Max Dessoir, den wir um seine Ansicht fragten. Was der bekannte Gelehrte und Schriftsteller über das Pferd schreibt, klingt sehr ruhig und etwas skeptisch, wenn freilich auch er eine bewußte Täuschung für völlig ausgeschlossen hält. Der Vorschlag, den er zur Klärung der Frage macht, erscheint uns sehr bemerkenswert. Wir geben nun den beiden Kritikern das Wort; zuerst dem Gläubigen, aus dessen Bemerkungen wir nur einiges bereits früher Gesagte und einige etwas allzu weit gehende Folgerungen gestrichen haben:

vor Liaojang.

Das Pferd des Herrn v. Osten.

Zwei Urteile. [Nachdruck verboten.]

Das Interesse für den „klugen Hans" in der Griebenow-straße nimmt, wie zahlreiche Zuschriften aus unserem Leser-kreise beweisen, geradezu beängstigende Dimensionen an. Es wird Zeit, daß man endlich einmal fernstehende und kalt urteilende Personen mit klaren Worten aussprechen, was eigentlich an dem „Wunder" ist. Das kann aber natürlich nur einen Zweck haben, wenn die Beurteiler das Pferd wiederholt ge-sehen und beobachtet haben; denn mit dem gering-schätzigen Achselzucken von Sportsmännern und Kunst-reitern, die den „klugen Hans" nicht kennen, ist es in diesem Falle nicht getan: diesen Fachleuten stehen doch zu viele ehrenwerte Männer gegenüber, die sich auch Fachleute nennen dürfen, und die auf Grund eigener Anschauung eine ganz andere Ansicht von der Sache gewonnen haben. Von dem Widerstreit der Ansichten geben zwei kleine Aufsätze Zeugnis, die wir hier veröffentlichen wollen. Der Verfasser des einen, ein junger Berliner Gelehrter, ist völlig überzeugt und sieht die Taten des „klugen Hans" bereits für den Beginn einer neuen Aera wissenschaftlicher Forschung an. Der andere Beobachter ist Professor Dr. Max Dessoir, den wir um seine Ansicht fragten. Was der bekannte Gelehrte und Schriftsteller über das Pferd schreibt, klingt sehr ruhig und etwas skeptisch, wenn freilich auch er eine bewußte Täuschung für völlig ausgeschlossen hält. Der Vorschlag, den er zur Klärung der Frage macht, erscheint uns sehr bemerkenswert. Wir geben nun den beiden Kritikern das Wort; zuerst dem Gläubigen, aus dessen Bemerkungen wir nur einiges bereits früher Gesagte und einige etwas allzu weit gehende Folgerungen gestrichen haben:

Neues von Hans.

Der kluge Hengst des Herrn v. Osten beschäftigt fortgesetzt die Aufmerksamkeit weiterer Kreise. Und mit vollem Recht: sehen wir doch erst am Anfang dessen, was hier ein gewichtiges und schwerwiegendes in unsere Naturanschauung eintritt...

Front Page News. *On August 29, 1904 the lower third of this page described "The Horse of Mr. von Osten." The enlargement shows only the first column.*

A circus newspaper published the results of a poll about the horse, concluding that Clever Hans was "the great sea serpent" that annually enlivened the columns of summer newspapers. The editor quoted the views of a foremost hippologist and was particularly concerned that circus performers everywhere had not banded together to oppose this phenomenon, which of course could be nothing more than small, learned routines.

Still others made offers to buy the horse or rent him for a show, knowing that Clever Hans was good business either way, so much so that masses of policemen were sometimes needed to keep the Griebenow crowds under control. One vaudeville company offered sixty thousand marks per month, but all offers were promptly refused. The horse was not available at any price.

Sea serpent or otherwise, Clever Hans helped the dailies that summer. The *Berliner Morgenpost* sagaciously predicted that his influence would last well beyond that year: "This thinking horse is going to give men of science a great deal to think about for a long time to come." And that was a correct prediction, foreshadowing the Hans legacy.

So adamant was the horse's owner, and so insulted by suggestions of fraud, that he finally appealed to the Board of Education in Berlin. But the Board was unsure of its own position in the extraordinary affair. Finally, it decided that a formal investigation should be conducted.

A committee was formed, principally to act as a court of honor for the two gentlemen most commonly accused of deception, Messrs. von Osten and Schillings. By early September 1904, their integrity was so much in doubt that the committee set itself the task of discovering "whether or not there is involved in the feats of the horse of Mr. von Osten anything in the nature of tricks, that is, intentional influence or aid, on the part of the questioner."

This committee, probably one of the most motley in German history, included a count, a physiology professor, the manager of the Berlin circus, a physician, two army officers, a former member of the city school board, a doctor of veterinary medicine, the director of the Berlin zoo and his assistant, a teacher in the public schools, a member of the Academy of Sciences, and a distinguished horseback rider. These thirteen august persons, all with reputations in fields related to the issue of Clever Hans, would settle the matter once and for all.

Leading this group was the man from the Academy of Sciences, *Sensory* Professor Carl Stumpf, Director of the Psychological Institute at the *Psychology* University of Berlin and the initial appointee to this position. As one of psychology's founders, Stumpf was interested in *sensory psychology*, which is concerned with the relationships between our

Carl Stumpf. *A few years after his study of Clever Hans he became rector of the University of Berlin.*

sense organs and experience. The focus is on seeing, hearing, tasting, and so forth. Sensory psychology was the first area of systematic research in the new science, and perhaps it was related to the problem of Clever Hans. Since all our knowledge comes initially through the senses, intelligence in the first place depends upon sensory impressions.

Earlier, prompted by a strong interest in music, Stumpf began an extensive examination of the psychology of hearing. When he needed special equipment, he purchased a neglected piano, pulled it apart, and made from the pieces a series of tuning forks carefully graded for different frequencies. With this apparatus he could measure a person's responsiveness to pure tones, and he achieved a foremost reputation in acoustics. But he kept these instruments in a cigar box, which does not qualify as an early psychology laboratory. So for several reasons, the honor of being the first modern psychologist has gone to Wilhelm Wundt.

But the leadership of this committee was appropriately entrusted to Stumpf. In his study of various sounds he had already demonstrated several relationships between a given *stimulus*, which initiates activity in an organism, and the *response*, which is the ensuing behavior of the individual. The question of Clever Hans would be approached in this same way. What was the stimulus that prompted the horse's correct response? Did Clever Hans really understand human language?

In approaching this problem, the eminent professor suggested that all members of the committee make direct observations of the horse. The first step, he emphasized, was to ensure that Clever Hans really could do the tasks.

After some difficulty Mr. von Osten was included among the questioners. Earlier he had declined this role, insisting that he would clear his name by remaining apart from all examinations of the animal, but when a member of the Commission one day announced that he had discovered the tricks involved, Mr. von Osten reversed his position entirely. He decisively agreed to participate with "any precautionary measure you may care to take." An emotional, unpredictable man, he fiercely objected to any attack upon the horse, which had responded successfully to many different persons. How could they all be involved in some fraudulent scheme?

The first meeting of the Hans Commission, as it came to be called, lasted over four hours, as did the deliberations of the following day. The usual conditions were established, questions were asked of the horse, and careful observations were made by all in attendance. These questions concerned the day of the week, the previous day, the next one, and various numerical problems. A local schoolteacher, Mr. Hahn, then added some new tests, all of which Clever Hans answered successfully. The horse was capable of most of the feats attributed to it.

There remained the perplexing problem of whether tricks of any sort were used, and in this quest the circus manager, Mr. Paul Busch, was especially important. He was given total freedom in making observations, examining the horse and its master in any way he wished. Other members were assigned to specific parts of Mr. von Osten's body, observing his head, hands, eyes, or other means by which he might signal the horse.

These tests involved colored cloths, photographs of persons in the audience, and certain words. It greatly impressed Mr. Busch when Clever Hans spelled his surname correctly. Hans was again successful, and in the conference that followed, wherein committee members exchanged information, it was found that no signs of trickery had been detected. Mr. Busch declared there was no possibility of a trick of any sort.

The next day Mr. von Osten agreed to a different procedure. He would stand behind the questioner, back to back, yet bending forward himself. In this way he would be effectively hidden from Clever Hans, but by his voice he could make the animal aware of his presence. The assumption here was that the horse was more likely to answer if he knew his master was present, yet the possibility of some visual sign was essentially eliminated. Despite this precaution and the uncomfortable posture for Mr. von Osten, nothing was discovered; the horse answered with very few errors.

Six slates were then suspended in a row, each with the name of a Commission member written upon it. Mr. von Osten pointed to one of the members and asked: "On which of the slates is this gentleman's name to be found?" Hans tapped the correct answer, and when asked to approach the slate, he did so, though without uniform success in every instance. These feats, declared Mr. Busch, were inconceivable.

The committee then asked for testimonial evidence. Count Otto zu Castell-Rüdenhausen emphasized that in eight days he had obtained dozens of correct responses, and Mr. Hahn gave similar support.

Later that afternoon, on September 12, 1904, the final day of deliberations, the members of the Hans Commission unanimously agreed that intentional signs were out of the question. They stated that despite the most attentive observation, none had been discovered and that "unintentional signs of the kind with which we are at present familiar are likewise excluded."

Several members emphasized Mr. von Osten's special teaching methods, which were highly systematic and ingenious. They had little in common with the usual training of animals, resembling instead certain schoolroom instruction. The report concluded: "This is a case which appears in principle to differ from any hitherto discovered. . . ."

So Mr. von Osten was temporarily cleared of fraud. But how then did the horse do it?

CHAPTER THREE

Background

L et it be emphasized that the Hans Commission was quite correct. There was absolutely no deception on Mr. von Osten's
part. It was exclusively through his teachings that the horse had
learned to perform successfully, and now the old man had a distinguished group of experts to support him.

This panel emphasized Mr. von Osten's methods of instruction,
which showed extraordinary ingenuity for that day. Perhaps therein
lay the answer. A proud, self-assured teacher with many years in
the classroom, he was also a great lover of horses and experienced
in their ways. Living alone in a fifth-floor apartment facing the
courtyard, he had considerable time to spend on his favorite project
and, most important, was fully convinced he would succeed.

But why was Mr. von Osten so convinced and so devoted? What
had induced him to take up this challenge in the first place?

Seemingly more attached to the beast than to his own kind, he
was influenced by recent ideas in biology and education, especially
Darwinian theory. If human beings have animal ancestors, he reasoned, the origins of human characteristics must lie in the animal

kingdom. Human beings bare their teeth in anger, but this action serves no useful purpose. In animals it readies for use a most powerful fighting weapon. Perhaps this human response reflects our animal nature.

Many people besides Mr. von Osten believed that the roots of human characteristics might be found in the lower species. On this basis animals might be a good deal more intelligent that previously thought. What had to be done, of course, was to train the animal's intellect a great deal more carefully. A mind not so imbued with evolutionary theory as Mr. von Osten's probably would not have undertaken this exceedingly difficult task.

In addition, a spirit of a different genre undoubtedly influenced Mr. von Osten. A philosophy that dates back to Aristotle had been revitalized by another Englishman, John Locke, and it came to be known as the *tabula rasa* viewpoint, which literally means "blank tablet." It states that the human mind at birth is like a sheet of paper with nothing on it. All the ideas and other responses contained in the human brain are exclusively the result of experience after birth. While not a startling viewpoint today, considering our general emphasis on the environment in human development, it earlier played an important role in prompting improved methods of teaching.

Introduction to Behaviorism   In this respect, Mr. von Osten was ahead of his times, for his methods with the horse illustrate a widely recognized system of modern psychology reflecting this view. Under the leadership of B. F. Skinner, this system also gained considerable popularity among social scientists, teachers, therapists, business people, parents, coaches, and others concerned with the control of behavior. Sometimes known as Skinnerian psychology, owing to Skinner's role in its development, it is more commonly called *behaviorism* because it focuses on overt behavior, rather than thoughts and feelings, which cannot be observed directly.

In particular, there is in behaviorism great interest in *learning*, the process by which we acquire and discard behavior patterns. Learning is any relatively permanent modification of behavior not based on some disability, and of course it was Mr. von Osten's chief concern with the horse.

To enable Hans to answer questions, gestures seemed to be the most obvious system. Hans might identify objects by pointing with his nose or picking them up in his mouth.

For numbers and words, Hans would tap with his foot. For words he would also use a special chart containing all the letters of the German alphabet in numbered rows and columns. To indicate any letter, Hans would learn to tap a pair of numbers, showing the row and column where the letter was located. Hans would indicate *i*, for example, by tapping three times, designating the row, and then twice, showing the position in that row. By tapping several pairs of numbers, the animal could spell a word, and anyone might know what he intended.

Teacher and Pupil. *The abacus, pins, balls, and chart all required special instructions.*

Principle of
Reinforcement   A teacher with years of experience, Mr. von Osten appreciated the importance of motivation in his pupil. Think of all the bored schoolchildren deficient in their sums! The horse, if it was going to master school subjects, would have to be motivated. So Mr. von Osten always carried in his back pocket a sweet carrot or some bread for use at the right moment, but Hans was never whipped or punished in any way.

Today in behaviorism this use of food, a compliment, a high grade, or similar outcome is called *reinforcement* because it strengthens or reinforces the behavior that precedes it. If a horse receives a carrot after making a certain response, that response is likely to be repeated. It will occur more often and more correctly as the reinforcement continues. Sometimes reinforcement is said to be synonymous with reward and punishment, but one never knows in advance how certain events will be regarded by an individual. Hence, reinforcement is simply defined as any event that increases the probability that a certain response will occur.

Responses followed by satisfying events tend to be repeated; those followed by aversive events tend to be discarded. Any behavior appears, reappears, or is modified on the basis of what happens immediately afterward. This form of learning is also known as *operant conditioning,* meaning that the organism operates on the environment to obtain some reinforcement.

Professor Möbius, director of the Prussian Natural History Museum, visited the horse and commented on this aspect of his education. "Mr. von Osten has succeeded in training Hans by cultivating in him a desire for delicacies. This desire is aroused by questions according to which the stallion acts . . . for as soon as he puts his foot down he snaps for the delicacy in the hand of his master." But the eminent zoologist remained skeptical about the animal's interest in school, apart from the delicacies. "I doubt whether the horse really takes pleasure in his studies," he concluded.

Method of
Approximation   In establishing communication with the horse, Mr. von Osten started by teaching Hans to point. This task would be easier than tapping, and the first step required almost nothing of the horse. Mr.

von Osten gave a command, pushed Hans' head himself, and offered a carrot. After many repetitions of this sort the horse eventually moved his head when it was merely touched by the trainer, who always spoke the command first and presented a carrot afterward. Still later, Mr. von Osten merely motioned in a certain direction when he gave the command, and Hans moved his head. And finally, the command alone sufficed.

It should be noted that Hans learned this pointing behavior very well, except for the directions left and right, which he usually confused. If the command was "left," for some reason he typically turned to the right and vice versa. This direction was to the left of the questioner, however, and Mr. von Osten explained, with abiding faith in his pupil, that Hans had the capacity to take the other person's point of view.

Hans then learned the tapping response by this same gradual method. At first his hoof was moved by the trainer; later it was merely touched; still later pointing was sufficient; and finally he lifted it on command, in anticipation of the carrot.

Then, to induce the horse to tap, rather than just lift the hoof, the trainer employed hand motions with the command. He motioned and spoke simultaneously, and at length the horse tapped on the command alone. If two taps were required, the commands were: "One, two."

This tapping response, once acquired, became an elaborate ritual. Hans stretched out his right leg in preparation, and for high numbers he showed his intelligence by tapping rapidly at first and then more slowly. For all numbers he always tapped at a certain distance from the original position of his front foot.

Later Mr. von Osten made this response more difficult, sometimes requiring the animal to "mark" a very large number by pulling his foot slightly backward, touching another place on the ground. He also introduced rest periods when several numbers were required, and on rare occasions he induced Hans to use the left foot, all of which made the usual tapping task more complicated for the poor animal. The stubbornness of the man exceeded even that of his beast.

Tapping Response. *When the correct number was reached, Hans swung his foot in a circular motion and returned it to its original position.*

In teaching Hans to tap and point, simple tasks were used first, gradually replaced by more difficult ones. This approach is called the *method of approximation,* for the steps are gradual; little by little the organism's behavior is molded in the desired direction. Thus, the old man, years ago, employed both of the basic principles of modern behaviorism. He used immediate reinforcement and the method of approximation.

If all this seems to have involved a great deal of effort, the reader should be informed that Mr. von Osten, a decade earlier, had spent three years training a previous horse, Hans I, which unfortunately died before his master's labors could be fully rewarded. But Mr. von Osten was not discouraged. He found signs of memory and pondering in that horse, expressions of a trainable mind. "If my horse makes a big arc in the street in order to turn into a gateway on the other side," he explained, "then that proves a consideration—namely, an autonomous thought process which is capable of being trained, if only I can make my intention understandable to the animal."

So enamored was Mr. von Osten with Hans I that he preserved the animal's skull after the poor beast had succumbed to a disease of the intestinal tract. He still believed in *phrenology*, much in vogue a half-century earlier, which stated that a person's character and behavior could be understood by studying the shape of the head. Applying the idea to animals, Mr. von Osten decided that Hans I had many promising lumps and contours and laid particular stress on the frontal structure of the cranium, significant for the intellect. He held up the skull of this dead horse beside that of the living one to emphasize their respective merits. The idea of phrenology had not survived earlier scrutiny, however; it was already an outmoded doctrine and had disappeared as the study of anatomy advanced. Too much was claimed, too little demonstrated.

But Mr. von Osten, with his zeal for such ideas and unrestrained confidence in the horse, was ready to develop his new Hans' intellect. Learning simple habits was one matter; learning to solve abstract problems would be quite another. Could he teach the horse, now proficient in tapping and pointing, to think for itself?

So the old schoolmaster continued, wearing his broad-brimmed hat and long duster in all kinds of weather. Tyrannical one moment, gentle the next, he was fanatically devoted to this cause, imagining all sorts of emotional tendencies in his animal and convinced that Hans engaged in subvocal speech.

Mr. von Osten was a practical man too, realizing that teaching aids would be essential if the horse were to succeed. He therefore collected some large wooden pins, some smaller ones, a series of colored cloths, placards with numbers up to 100, and a counting machine. He also prepared the spelling and reading chart described earlier, and for music he obtained a child's organ with a diatonic scale. He was thus ready to teach pre-school and classroom subjects.

He started with counting, using much the same system that had earlier proven successful. He placed some duck pins on the ground and then, by commands and hand motions, elicited three taps when three pins were in front of the horse: "One!" "Two!" "Three!" Eventually after months of training, the numbers and then even the gestures became unnecessary. The question itself was sufficient to elicit the correct response. The pins were placed before the horse and the teacher simply asked: "How many pins are there?"

At times both the horse and master became discouraged, but eventually Hans became almost uniformly successful with the basic digits. For zero he even learned to shake his head from side to side, for there was no other convenient method to indicate that no taps were required. He always shook his head in the same way, turning first to the right, then the left, and never in the opposite order. This response also indicated "No," as when Hans was asked how many corners in a circle.

Our system of counting is not entirely logical, as any child discovers when corrected for saying "nine, ten, one-teen, two-teen." But Mr. von Osten, adhering to the method of approximation, began with these expressions. After reaching ten, the horse was taught "einzehn" and "zweizehn," meaning "one-teen" and "two-teen" in German. Later, when the animal was ready, these words were replaced by "elf" and "zwölf," meaning eleven and twelve.

Mr. von Osten, in ingenious fashion, used all types of approximations and reinforcement. To teach the concept "and" in mathematics, a large cloth was held in front of the horse and slowly raised while the word "and" was pronounced in careful tones. The word became associated with the cloth, and when the pins became visible, Hans naturally tapped out their number, for which he was rewarded. Later, some pins were placed in the open or on work tables, and others were hidden under the cloth. The horse tapped out the first part of the answer, the pins in sight, and then, as the cloth was raised and Mr. von Osten said "and," he tapped out the others. Mr. von Osten's great satisfaction at the first success here hopefully did not go unnoticed by the horse.

Eventually it was possible to omit the cloth entirely. Hans was simply shown two sets of pins and asked, "How much is two and four?" After countless trials the horse invariably gave the right answer and, after much further work, even the pins could be omitted. Hans had learned to manage numbers in the abstract, without counting and without the objects in front of him.

Work Tables. *In teaching addition, Mr. von Osten placed some pins on each table and asked the horse the sum.*

Mr. von Osten emphasized that Hans might have been taught small sums in rote fashion, such as two plus two, three plus three, and so forth, but when it came to two-digit numbers, multiplication, and division, all of which Hans eventually accomplished, mere practice was impossible. That would require far more than a lifetime. The old schoolmaster was satisfied that Hans had learned to think for himself, having mastered the fundamental mathematical processes.

Behavioral Prompts   Some of these teaching aids were used to present a task to the horse. The child's organ sounded different tones for examining musical ability, and the cloths were used to test color discrimination. Other aids helped the horse solve the problem. The large cloth placed over some pins and even the pins themselves made addition and substraction easier, and the counting machine simplified multiplication and division.

This simplification of the problem through special apparatus, especially in the early stages, is another technique in behaviorism. When some device serves in this way, to encourage the desired response, it is called a *behavioral prompt*, for it increases the learner's chances of success. To learn to swim, a child uses a training board; to walk alone, an elderly person uses a cane; and to teach the word "times" in multiplication, Mr. von Osten used the counting machine. All are behavioral prompts.

This machine contained several rods, each with ten balls strung along it. Two of the balls were pushed far to the left and the horse was asked: "How many are there?" The horse tapped twice and the trainer answered: "Very well. That is once two." Two more balls were pushed to the left but a short space separated them from the first pair. "How many times two balls are there?" As the trainer asked this question, he made a pronounced hand movement toward the two separate groups. The horse responded with two taps. "How many, therefore, are two times two?" The horse, with the four balls in view, answered with four taps.

The space between the pairs, like the cloth in learning "and," helped the horse learn the meaning of "times." Both were behavioral prompts. The horse was taught to notice and count groups by the spatial separation, and no questioner was allowed to use the command "multiply." Hans was always asked to take one number times another. Mr. von Osten never deviated from this practice.

Four years of such instruction seem almost impossible, to say nothing of the earlier horse, but testimony to Mr. von Osten's efforts came from many sources and thousands of spectators. Major von Keller, who had known him for 15 years, supplied a written statement, and General Zobel indicated that he had become acquainted with these procedures fully a year before the first public exhibitions. Neighbors said that for years the master and horse had behaved like a teacher and child in school. The whole plan was beautifully conceived and executed, reflecting the most advanced methods in classroom instruction.

This approach today would be called *behavior modification,* which is the application of behavioristic principles for solving some practical problem. When used in therapy and child rearing, the aim is to eliminate undesirable habits and to produce more desirable ones, but behavior modification is also employed in a wide variety of work and school settings. Here the aim is to teach all sorts of knowledge and skills, which was Mr. von Osten's fundamental purpose with the celebrated horse.

Behavior Modification

As a formal approach to the teaching-learning situation, behavior modification was unknown in Mr. von Osten's time. But apparently he discovered several of the basic principles himself, perhaps as a result of his many years in the classroom.

And as Clever Hans began to give solutions to new problems, ones quite different from those on which he had been trained, Mr. von Osten found the proof he had so dearly sought. The animal was capable of independent thought. The horse was proving himself intelligent, not just trainable, and lack of speech was not an insurmountable barrier. The crucial factors were an innate predisposition and careful training, and here Mr. von Osten's extensive instruction in mathematics was a wise choice. In practically no other subject can concepts be so readily illustrated by behavioral prompts and the evidence of success or failure so readily demonstrated.

45

By 1904, Mr. von Osten had spent thousands of hours training his second Hans, and for all this work he received more publicity than he wanted. The newspapers called him Professor von Osten and Baron von Osten, and he regretted this premature fanfare. The project was done purely from a scientific standpoint, he said, and it could be successfully repeated with any horse of average ability.

CHAPTER FOUR

Forming Hypotheses

With the failure of the Hans Commission to find any explanation for the horse, Professor Stumpf decided that another investigation was necessary. He recommended an extensive series of tests and chose as the principal figure for this work a young man named Oskar Pfungst, a graduate student in his laboratory at the University. Pfungst was charged with answering the perplexing question of how the horse did it, and he is the person to whom we are chiefly indebted for this tale of Clever Hans.

With his background in sensory psychology, Pfungst probably had some specific hypothesis in mind; perhaps he was on the lookout for auditory or other sensory cues, which is why Stumpf appointed him in the first place. We do not know for certain, but we do know that most researchers approach a problem with some sort of provisional explanation or guess, to be held until further evidence displaces it. We know too that among the diverse hypotheses in the case of Clever Hans, some were based on scientific research; some were derived from theory; and still others were pure speculation, arising from personal interests. An examination of these possibilities gives some idea of the thinking at the time.

Discovery of N Rays One explanation for Clever Hans came from the best scientific laboratories in the world. It was stimulated by the discovery in 1895 of X rays, so-called because their nature was so puzzling and so totally unknown previously. Within a year dozens of books and more than one thousand articles appeared on the subject, and in March 1903, while Mr. von Osten was reaping his unusual reward in the Griebenow courtyard, a distinguished French physicist, Professor Rene Blondlot, contributed another startling finding. He came upon a new form of radiation, which he called *N rays* in honor of his city and institution, the University of Nancy. These

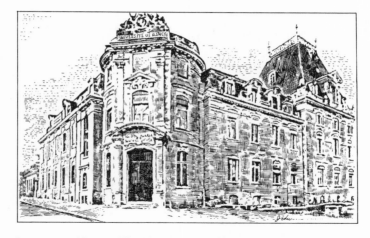

Institute at Nancy. *The University was founded at Pont-a-Mousson four centuries earlier and later was transferred to Nancy, where this physics building was constructed.*

radiations were also invisible, but with prisms they could be separated or concentrated in different locations. He therefore knew that he was not observing X rays, as he emphasized to the members of the Academe des Sciences.

Subsequent studies showed that N rays came from certain metals, especially steel, and also from the sun. Encouraged, Blondlot inaugurated a large-scale research program, and others joined the investigation. The French physicists were particularly vigorous; Mace de Lipinay, Guitton, Bichat, and Guilloz all made scientific reports, demonstrating that N rays were stored and emitted by quartz, carbon sulphide, and even limestone, while negative results were obtained from wood, paper, and aluminum.

N rays of course were invisible to the naked eye in daylight, but they were readily observed in darkness. N rays flowing from a piece of metal increased the brightness of an electric spark, and an aluminum prism, which did not emit this form of radiation, could be used to bend it.

The following year brought forth more reports and a most important discovery by Augustin Charpentier in the Department of Biophysics at Nancy. He found that a similar radiation was fundamentally involved in biology. These variant N-type radiations were emitted by animals and human beings, as well as certain plant tissues. Most important, they were concentrated chiefly in contracted muscles and activated nerves in the body.

Using a sensitive screen, Charpentier developed a technique for detecting bodily movements and simple mental exertion via N rays, as reflected in the activity of the nervous system. Others demonstrated that N rays increased our basic sensory capacities, including seeing, hearing, smelling, and tasting, and this research became legion all over Europe from Belgium to Italy.

With Clever Hans performing at just this moment and on such a puzzling basis, it was hypothesized by some that the inexplicable N rays or variant N-type radiations were responsible. They were transmitted from Mr. von Osten to the horse or emanated spontaneously from the animal. This hypothesis had a most prestigious source, and it seemed all the more likely because X rays, just a few years earlier, had proven so puzzling and significant to humanity. Prior research is a most important ground for scientific hypotheses, and so it was with Clever Hans.

Besides research findings, hypotheses are also stimulated by theory. A *theory* is a set of related principles and premises with explanatory value; it is a way of making sense out of some facts, providing an interpretation. From this general interpretation, specific hy- *Influence of Theory* potheses can be derived and tested, each on its own merits. A good theory therefore stimulates and guides research.

In the case of Clever Hans, evolutionary theory was most influential. Charles Darwin's work had given some people the idea that the horse, somehow through natural selection, had acquired an extraordinary mental potential. This capacity had lain fallow until it was developed by Mr. von Osten, though perhaps it existed in other horses and even other species. Mr. von Osten, from all his work, knew that he had evoked a latent intelligence. That was the only answer.

Charles Darwin. *Before his trip a new microscope was among his most important possessions.*

The Journal. *On arrival home Darwin had a new manuscript, his journal of the voyage.*

Certain zoologists also adopted this view. They saw in Mr. von Osten's work a demonstration of the continuity between the animal and the human mind, a doctrine that had been developing steadily since Darwin. But careful, systematic educational methods heretofore had not been tried extensively with an animal. In fact, Dr. Ludwig Heck, Director of the Berlin Zoo, declared that Mr. von Osten had just completed one of the most remarkable demonstrations in human history.

Pet lovers agreed. Here was evidence that animals, in certain respects, are perhaps more intelligent than human beings. Homing behavior in pigeons, altruism in porpoises, and dogs' knowledge of impending storms were emphasized. The study of Clever Hans was revealing enormous capabilities in animal mental life.

Other enthusiastic souls, well aware of Darwinian theory but not knowing quite what to make of the horse, instead made fun of the whole situation. There arose all sorts of Clever Hans jokes and bemused speculation about the future of horses. "Many a young lieutenant," wrote one observer, "will be embarrassed to put his spurs to the nag which can add better than he." "We humans," he warned, "who put so much stock in knowledge and progress, will do well to pack up our wisdom and with every passing coach horse doff our hats respectfully. Who can say whether or not some secret Socrates lies inside that melancholy skull?"

Among those acquainted with the works of Jonathan Swift there was the inevitable reference to Gulliver's lesser-known travels in Houhyhnms Land. Pronounced "whinneums," as in the horse's whinney, this country is ruled by highly intelligent horses who communicate exceedingly well with one another. Their beasts of burden are irrational creatures called Yahoos, having no tail and little body hair, though the male Yahoo sometimes has a long beard like a goat. In both sexes a thick head of hair appears in any one of several colors, including yellow, brown, black, and red. To his dismay, Gulliver notices that the physical structure of the Yahoo closely resembles his own, and suddenly he realizes that Yahoos are people and he too is a Yahoo. Houhyhnms Land is the world of horses and human beings turned upside down, just as it sometimes seemed in Berlin.

Telepathy   In addition to research and theory, hypotheses also come from
Hypotheses   personal inclination. Private interests and attitudes determine what we choose to study, and the case of Clever Hans was no exception. Many people, through personal experience, were prompted to a hypothesis that attracts broad interest today, concerning the possibility of extrasensory perception.

The stem *extra* means "outside of"; *extrasensory perception* or ESP is perception that occurs outside of our normal senses. It is the alleged ability to know about things without hearing, seeing, touching, tasting, or smelling them, or using any other known sensory capacity. There are several alleged dimensions of ESP, but one that enjoyed the best standing in Hans' day and still captures the imagination is *telepathy*, in which one person knows another's thoughts without assistance from any known, normal sensory channel.

The telepathy hypothesis held considerable weight in some quarters, and some years earlier it had been the primary concern of an international research group in London, the Society for Psychical Research. This society explored all kinds of psychic phenomena, from post-mortem survival to water dowsing, and by the time of Clever Hans two views of telepathy had emerged. According to one, telepathy went from brain to brain like a wireless telegraph. Some physical medium was involved and the process followed the laws of our physical world, however poorly understood. The other view was also speculative. It regarded telepathy as a mind-to-mind connection, following the unknown laws of the psychical world. A corollary of this view is that the mind can exist apart from the body, even after bodily death, an idea sometimes referred to as post-mortem survival.

There was little possibility of testing these views in the Clever Hans case, but many people believed in this explanation. The horse simply read the mind of his master and others as well and therefore was widely known as "the telepathic horse." Occasionally, he even answered a question before it was asked, which emphasized his mind-reading ability.

Extrasensory perception includes two other dimensions, one of which complicates the telepathy hypothesis. In *clairvoyance* someone has knowledge of some current event, and in *precognition* someone has knowledge of a future happening, all without any known sensory awareness. On this basis it is sometimes difficult to devise a pure, extended test of telepathy. If there is any written record of the thought, or if the event being thought about existed previously, the successful psychic, so the argument goes, might apprehend it by clairvoyance. Because of this difficulty, telepathy and clairvoyance have been studied together, and then they are referred to as GESP, meaning general extrasensory perception.

A series of tests with Mr. Schillings seemed to provide evidence for pure telepathy, nevertheless. This famous African adventurer asked someone to concentrate on any number between one and twenty, without writing it down or disclosing it to anyone. Then Hans dumbfounded all those in attendance by tapping out the number, known only to that person and not even to Mr. Schillings himself.

In fact, Hans sometimes was so clever at knowing answers that seemed impossible, such as the age of some member of the audience, that Mr. von Osten cautioned spectators not to whisper their thoughts. If the horse heard the answer, or if he read someone's mind, it certainly was not a good test of his ability to think for himself.

Suspicion   Finally, despite the findings of the Commission, many people
of Trickery held to the idea of trickery. The answer to the clever horse lay with
his suspicious-looking owner. The horse's feats, if he did them at
all, were the product of some subtle deception by Mr. von Osten or
his companion in this misdemeanor, Mr. Schillings. But why had
Schillings, a popular figure and successful in his own right, fallen
into the category of deceiver?

All kinds of tricks were suggested. Mr. von Osten signaled the
horse by using his slouched hat; there was something under his
great coat; Schillings did it with eye movements; underground wires
led to the horse; or people in the audience sent signals.

There were successful questioners who wore neither floppy hats
nor long dusters, and the horse performed successfully in Schilling's
absence. But as Charles Darwin said, evidence that runs contrary to
our convictions is often ignored. Tricksters, moreover, are never in
short supply, as attested by several personalities at the time.

One such figure was Madame Eusapia Palladino, a self-pro-
claimed Sicilian psychic who performed all over Europe, demonstrating how easily the public is fooled. In any small room in semi-
darkness, where she always held her sittings, objects floated from a
cabinet; tones came mysteriously from a guitar; the curtains bulged;
and a table rose in the air. Members of the audience heard spiritual
voices and indeed were touched by the spirits. Eusapia Palladino
apparently was making things happen without any direct action on
her part, a capacity now called *psychokinesis* or PK, which literally
means mind-movement or mind over matter.

The difference between psychokinesis and extrasensory perception is readily understood in terms of elementary physiology. One
part of the human nervous system is sensory, given over to receiving information about the world, and the issue here is ESP. Another portion is motor or muscular, devoted to doing something in
the world, and the psychic claim here is PK. In short, ESP involves
receiving signals, and PK involves sending them. This difference is
illustrated by a friend in downtown Boston. He says he finds parking spaces by ESP. It might be quicker if he could create parking
spaces by PK, but of course that would not be courteous to other
drivers.

ESP and PK constitute the two basic dimensions of alleged psychic experience. Together they are referred to by the term *psi*, pronounced with a heavy heart, like "sigh." It means any ability or event that cannot be explained on the basis of our presently known natural laws. Most behavioral scientists do not accept the idea of psi, but when they do, their investigations are known as *parapsychology*, meaning the study of psychological phenomena which do not seem to follow our currently known natural laws. Sometimes these investigations are also referred to as psychic research.

Instead, behavioral scientists typically look for simpler explanations, focusing upon such possibilities as trickery, forgetfulness, or the unusual use of our normal abilities. But several highly respected scientists inspected the work of Eusapia Palladino and found it above reproach.

She eventually arrived in the United States, and not long afterward a good deal of pressure was put on Hugo Munsterberg, a psychologist at Harvard, to scrutinize her work. He had refused all such activity in similar instances, claiming that a man of science, trained to believe what he sees, was unsuited to the task. He suggested instead a professional magician but finally agreed to a sitting with Eusapia, saying: "Madame Palladino is your best case. She is the one woman who has convinced some world-famous men. I was never afraid of ghosts; let them come."

In preparation for Munsterberg's investigation, Madame Palladino placed an electrical burglar alarm in the window of the seance room, emphasizing that she received no outside help. During this performance, as in all others, she kept both hands on the table, and her feet were held in position by reliable members of the audience. Despite these procedures, one side, then the other, and finally the whole table lifted itself into the air. Mysterious rappings were heard. A sudden breeze blew from the cabinet. A member of the circle received a kiss—or a punch, it was difficult to tell which—on the cheek. These feats were accomplished with observers sitting right next to her, one on each side, with the person on her right grasping her right arm and touching her right foot with his own, while the person on the left was grasping and touching Eusapia's left side.

Madame Palladino. *Some people decided that she controlled her abdominal muscles like an Oriental dancer or concealed some special instrument beneath the folds of her skirt.*

Munsterberg, well versed in the scientific method, was prepared to use it. He decided that all conditions were well controlled except one, some wires probably attached to her body. So an accomplice hid under the table to watch for them.

In the midst of this particular seance, just after the mysterious rappings had been heard, there suddenly came a very wild scream, as though Eusapia had been stabbed with a dagger. But she had been nabbed, not stabbed, by Munsterberg's colleague just as she was reaching for the guitar and a small table in the cabinet—with her foot.

Her tricks, it was discovered, were accomplished in this fashion. While observers sat beside her, grasping her arms and touching her feet with their own, she wriggled out of their full control. Undetected, she freed her left leg by placing her right foot so that it touched those of both her observers, or else she simply extracted her left foot from its shoe and pressed down on the empty shoe by means of a hook attached to her right one. Then she proceeded, with incredible acrobatic movements of her left leg, to lift the table, kick the curtains, pluck the guitar, thump on the cabinet, and even pinch people through a curtain by using her toes.

Fraud with the feet! But still no mean feat. As Munsterberg said even before she was caught, Eusapia had to be "a woman with unusual skill, unusual talent, unusual strength. . . ." After each seance she was "in full perspiration" and for good reason. Hers was not easy work.

And that's the tale of two sittees, Munsterberg and his accomplice. They turned the table on the lady from Sicily.

She in turn, showed once again that the public is very ready to be deceived and that there arise daily those who would profit from this inclination. Even in our own times there is little risk in this business, for the exposed fraud often retains public interest.

Dr. Meissner, assistant at the Royal Veterinary College, added to the suspicions of trickery in the case of Clever Hans. Anyone who sees the horse's performance with open eyes, he said, should be able to recognize the cues or "helps" provided by the trainer. "After I watched this comedy for an hour, a feeling of embarrassment engulfed me—that educated people of the twentieth century from the highest social circles could not notice that they had in front of them horse training, the way it is practiced in every circus."

But the veterinarian's case was weak, for he did not say how the animal had been trained, and he did not indicate the type of "helps" that he suspected. He even suggested that Mr. von Osten was parading his horse for money, but the reader knows that there was no monetary advancement whatsoever for the elderly trainer.

Sensory Cues   We should add that Carl Stumpf, leader of the Hans Commission, was forced to be absent from Berlin from September 17 to October 3, during Pfungst's initial inquiry, but he had his own ideas on the subject and expressed them privately. One concerned the so-called nasal whisper, sometimes used to explain instances of alleged telepathy. A Danish scientist named Lehmann had suggested that organisms with acute auditory sensitivity could respond to the softest whisper, in which the thought was only spoken inwardly, with the lips tightly closed. This possibility seemed to Stumpf worthy of investigation; perhaps Hans was responding to minimal sounds.

This hypothesis was based on the concept of threshold, which concerns a very slight stimulus or change in stimulation. Specifically, the *absolute threshold* is that level of stimulation that can just barely be perceived by a given individual. It is the point, determined over a large number of trials, of just noticeable stimulation for that individual. By definition, an organism can experience stimulation only when it is at or above its absolute threshold.

According to Stumpf, the nasal whisper might be above the horse's absolute threshold, though out of range, or below threshold, for people in the audience. The answer to the riddle of Clever Hans perhaps lay in the different absolute thresholds for hearing in the two species.

Mr. von Osten also emphasized hearing. He demonstrated that if he closed both his nose and mouth, asking the question to himself, the horse made no successful response. If he closed only his nose, leaving the mouth open while inwardly stating the problem, Clever Hans could hear what was asked and thus answered a question

58

inaudible to human ears. The trainer emphasized his point by using a placard to deflect the unspoken sounds toward or away from the horse's ears, with the expected result in each case. On another occasion he put earmuffs on the animal, and then Hans could not answer the questions.

The ears were not the only sense organ made to bear some burden of explanation. The nose, too, was implicated by some, especially in the identification of a photograph of a member of the audience. That person carried the picture around with him, impregnating it with his own peculiar odor, which the horse perhaps detected. Stumpf placed little credence in the sense of smell, however, and also in vision, which might have been invoked in these instances. The domestic horse certainly could not recognize the details of a photograph in this fashion.

In public Stumpf was more reserved. Awaiting further tests, he refused to pass judgment of any sort on the intelligence of Clever Hans or the character of Mr. von Osten. Just as it would be unscientific to come to any decision about the horse without examining it, so it would be equally unwarranted to pass moral judgment on a human being with so little evidence. The responsibility for those who sit in judgment is to point to the proof.

Stumpf merely described the horse as the most astonishing creature "that can be seen in the animal realm of pedagogy." But he did not believe in highly developed conceptual thinking in animals. He added, "I am absolutely prepared to revise my opinion as soon as persuasive facts can be brought forward."

Altogether then, before Pfungst began his work, there were at least four prominent hypotheses for Clever Hans' behavior:

1. The newly discovered N rays were the responsible agent.

2. Certain animals have the capacity for extraordinary intelligence, if they are properly trained.

3. Clever Hans possessed some form of extrasensory perception, probably mental telepathy.

4. The whole performance was based on trickery by Mr. von Osten and perhaps some others, too.

According to Professor Carl Stumpf, the nasal whisper was still another possibility, but he did not promote this idea in any significant way. He mentioned it to Pfungst and apparently to no one else.

With these and his own ideas in mind, Oskar Pfungst erected a large tent in the courtyard. Afterwards, he began work.

CHAPTER FIVE

Testing Hypotheses

The purpose of the tent was to keep out distractions of all sorts, including inclement weather and the spectators who arrived at Griebenow Street in endless numbers. Pfungst later moved some of his work to the horse's stable, and at other times he made tests in the open area, depending upon which factors were under investigation.

Just two weeks before he began, however, there was suddenly more information on the N rays hypothesis. In the laboratory of biophysics at Nancy, people had been kept in dark closets for nine hours, entirely removed from any source of light, and no decrease in emanation of N rays was found when they emerged. In other words, N rays in human beings apparently were not the result of radiation from the sun. They seemed to be a basic characteristic of human functioning.

Successful photographs also had been made. This radiation could not be shown on the photographic plate, owing to its invisibility in normal light, but its effects were readily displayed in the dark. An electric spark, which increased in brightness when contacting the rays, could be seen in successive photographs and the results evaluated with quantitative methods. These measurements apparently ruled out subjective factors, for which this research had been criticized.

Most important was the work of an American physicist at Johns Hopkins University. During September 1904, Professor R. W. Wood visited a Brussels laboratory where N rays had been observed. Working necessarily in darkness with a Belgian scientist, Wood wished to know how accurately the new form of radiation could be detected by the unaided human eye. He secretly shielded the N rays emanating from a piece of metal and found that the Belgian did not report any decrease in brightness. Then he unobtrusively removed the prism used to deflect them and found that the Belgian did not report any straightening of the N rays. Later, he surreptitiously replaced a piece of steel from which they supposedly flowed with a block of wood of similar shape and size, widely known to be a nontransmitter, yet the Belgian continued to report N rays.

By these procedures he discovered that the variations in brightness that supposedly indicated N rays did not coincide with any physical event at all. Instead, the event was a psychological one, just the idea itself. The Belgian scientist and others who reported N rays were simply "seeing things." N rays existed only in their imagination.

Scientists are human beings; they too can be fooled. Their wishes and expectations can influence how they perceive things, and especially in France many scientists were motivated to see N rays. This new form of radiation was first detected there; it was observed there more often than anywhere else; it was more commonly measured there; and when the truth finally became known, resistance lasted longest in Nancy. These scientists, in particular, saw what they wanted to see, creating their own "evidence" for the new radiation.

With this result and similar findings elsewhere, reports of the remarkable new radiation rather abruptly ceased. Only a fortnight before Pfungst's research, N rays no longer seemed a suitable explanation for the feats of Clever Hans.

There remained the other possibilities, but before beginning his experimental work Pfungst wanted to confirm the findings of the Hans Commission. For this purpose he enlisted Messrs. von Osten and Schillings and some of the neighbors to act as questioners. He even served in this capacity himself, thinking that the horse might answer correctly for others but not for himself. These efforts yielded no evidence of tricks, however, and he had not suspected them anyway.

Courtyard Neighbors. *The Piehl family, gathered at the bottom of the stairs near their carpentry shop, often assisted with the horse.*

Pfungst's next step was to begin the experimental procedures, perhaps using some sensory hypothesis. In the *traditional experimental method* something is manipulated; observations are made while apparently relevant factors are changed in one way or another. Through this procedure the investigator discovers what influences what, which is the purpose of the experimental method. Each possibly significant factor is observed, manipulated, and then observed again, and the effect is noted. This work deserves our close attention.

At the outset Pfungst simply wanted to find out whether the answer really came from the horse, in which case Clever Hans was extraordinarily intelligent, or from somewhere else, in which case he only seemed to think like a human being. ESP, trickery, or some other factor would be the explanation in the latter case.

To study Clever Hans' intelligence, Pfungst established two different conditions for asking him a question. Sometimes the questioner knew the answer, such as a color, the time of day, or the number of straw hats in the audience. This procedure was called "with knowledge." On other occasions, called "without knowledge," the examiner asked a question to which he himself did not know the answer. Only the horse might know because the questioner either did not know what he was asking or did not know the answer. This manipulation of the questioner's knowledge was the essence of the experimental method at this point.

These two circumstances, with knowledge and without knowledge, are called the experimental and control conditions, respectively. The *experimental condition* includes the factor under investigation, which in this case is the examiner's knowledge of the answer. The *control condition* does not include this factor; it is omitted or varied in some way. In other words, the experimental and control conditions áre as identical as possible in all respects except for the factor under investigation. If a different result occurs—if the horse behaves differently from one condition to the other—it is presumed to be attributable to that one factor of difference between the two conditions.

In the decisive experiment on N rays, a piece of steel constituted the experimental condition, for it presumably emanated N rays. A block of wood, known to be a nontransmitter, was used for the control condition. Since the reports of N rays were the same under both conditions, the type of material, steel or wood, apparently made no difference.

With Clever Hans these conditions were established by using placards on which numbers were printed in large type, one to each card. For the experimental condition a placard was read by the examiner and then shown to the horse. In these "with knowledge" trials Hans readily tapped the correct answer.

For the control condition the placards were randomly scrambled face down and then, without seeing what was written on the other side, the questioner held one up for the horse. Here Hans' success decreased markedly. His score dropped from 98% correct to about 8%. He was successful occasionally, perhaps by guessing, but when the answer was unknown to the questioner, the horse was incapable of giving it himself. Apparently Hans could not read numbers.

To discover whether Hans could read words, a similar procedure was followed, except that each card had a simple name written on it. These were hung in such a way that the horse could see them, and then experimental and control conditions were used. Sometimes the questioner was with knowledge, having read the card, and sometimes without knowledge. In fourteen such tests the horse disgraced himself again. He was correct in 100% of the former instances and none of the latter. The animal could not read words, either.

Pfungst then spoke a number directly into the horse's ear and Mr. von Osten did so too. Afterward they asked the horse the sum of the two numbers, and this test constituted another "without knowledge" trial. Neither gentleman knew the number whispered by the other, and the answer could only come from the horse. With this procedure the horse again failed. It left Hans not knowing what the right Hans should be doing. In thirty-one control tests he was correct only three times, and in thirty-one experimental trials, in which the questioner knew the sum, he was correct twenty-nine times.

These experiments show Pfungst's ingenuity as an investigator, and they also illustrate the importance of controlled conditions, since he eliminated unwanted influences. In fact, to decrease the likelihood of a recording error, he had Dr. von Hornbostel write down the results from each test, leaving himself free to manipulate the apparatus and manage the horse. From time to time he also led Clever Hans around the courtyard or let the animal run free. He wanted to test the horse in its normal condition, not when it might succeed due to outside influences or make errors due to fatigue.

So Oskar Pfungst had demonstrated that Hans could not read, spell, or count unless the examiner knew the intended word. With calendar problems, tests of musical ability, and questions about telling time, the horse performed in the same way. He was correct on 90% of the with-knowledge trials, compared with 10% for without-knowledge tests. Taking the factor of chance into account, the latter result was unimpressive.

Naive examiners asked Hans the simplest question of all and found that he did not even know his own name. The horse's far-famed ability was an illusion.

As Pfungst wrote: "Hans can neither read, count, nor make calculations. He knows nothing of coins or cards, calendars or products, nor can he respond by tapping or otherwise, to a number spoken to him but a moment before." Hans could not think like a human being.

Of the four widely cited explanations, only telepathy was left, along with Stumpf's idea of the nasal whisper. N rays were fictitious; trickery was unlikely; and the animal had no special intelligence.

An observation in one series of tests seemed to support the telepathy hypothesis. On "without knowledge" trials concerning the days of the week, Hans answered four of the questions correctly, which was far above chance. Hans' groom, present during the testing, knew the answers, and according to some people the horse was reading the man's thoughts.

To test the importance of directly observing a knowledgeable questioner, a different procedure was planned, and it would have a bearing on telepathy. The questioner would know the answer in both conditions, but on the control trials Hans would not be permitted to observe this person. The two conditions would be "with observation" and "without observation," respectively. If the animal could respond correctly in both cases, telepathy was a possibility.

For these tests Pfungst prepared another bit of apparatus. He made a very large blinder and fitted it to the right side of the horse's head. But Hans did not like it. He reared into the air, thrashed about, and turned his head wildly from side to side. To construct a useful blinder, Pfungst eventually had to make it three or four times its normal size.

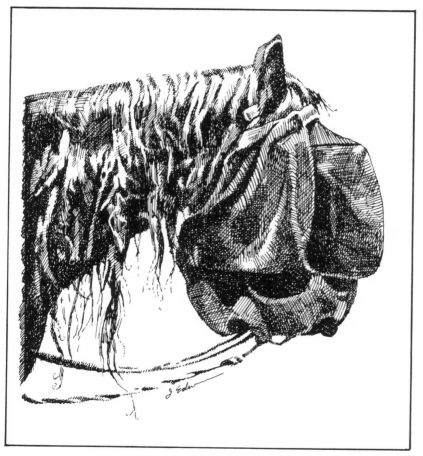

The Blinder. *Various devices were used in later tests. Here a large leather piece was supplemented with a dark cloth.*

While wearing the blinder, the horse's inclination to look was so strong that it resulted in an "undecided" category, in which the investigators were unsure whether Hans, in his strenuous efforts to free himself from the necessary restraints, had caught a glimpse of his questioner. Altogether, 108 trials were made, and they produced 82% correct answers in the experimental condition. In the control situation, which prevented direct observation, Hans success fell to 24%. The undecided category showed Hans to be correct 72% of the time, which suggests that he did a great deal of peeking.

Later, the questioner stood behind a screen, and the results were equally clear. If the horse was truly a mind-reader, he ought to be able to do so without looking at the person, but Hans was again inept.

Manipulation and Observation

In all of these tests, Pfungst was concerned with two significant types of factors or variables. An *independent variable* is any changeable element, usually some form of stimulation, manipulated by the investigator. It is independent of all other influences except the investigator's manipulations. Pfungst manipulated the questioner's knowledge, which was therefore the independent variable.

The *dependent variable*, usually some response, presumably changes with manipulations of the independent variable, providing all other significant factors are controlled. Giving a correct answer was the dependent variable in the study of Clever Hans, and it depended upon the horse's opportunity to observe a knowledgeable questioner.

In other words, the independent variable is present or manipulated in the experimental condition and absent or held constant in the control condition. In fact, its presence defines the experimental condition; its absence constitutes the control condition; and its influence is determined by comparing the two outcomes, when all other factors are held constant. So here is the underlying question in any traditional experiment. When the independent variable is manipulated, what happens?

Hugo Munsterberg's scrutiny of Madame Palladino illustrates this procedure, sometimes known as the *rule of one factor*. In the traditional experimental method only the independent variable is manipulated or varied in some way; all other significant factors are eliminated or held constant. Eusapia wanted to give this impression herself, placing the burglar alarm in the window, gentlemen at her sides, and her hands on the table. The single factor at work, she claimed, was her psychokinesis, which made the table rise, the curtains bulge, the guitar play, and other strange happenings, all presumably dependent variables. But when Munsterberg's accomplice exerted still another control, grabbing her leg, Eusapia's PK had no effect. Her alleged psychokinesis accounted for nothing at all.

Pfungst made similar efforts with Clever Hans. He erected the tent, eliminated spectators, dismissed the groom, and even asked Mr. von Osten to stand aside, all to achieve controlled conditions. Then he manipulated one factor at a time, the independent variable, and by this method discovered the key to the horse's success. The animal somehow was obtaining information from a knowledgeable questioner. But how?

Earlier Pfungst had noticed that shouts by the spectators did not interfere with Hans' performance. Those who boldly called out "Halt" or "Wrong" during the tapping process did not disrupt the horse, and there had been almost a total absence of ear movements during these tests, a curious condition in such an attentive and spirited horse. Hearing apparently was not an important factor. Mr. Henry Suermondt, the distinguished horseback rider and Commission member, supported Pfungst in this opinion, casting further doubt on the nasal whisper as the responsible agent.

Eventually, Pfungst noticed that when Hans was asked to count the straw hats in the audience, he looked at the spectators and then made a subtle effort to observe the questioner. In other tests his performance grew steadily worse as darkness approached and as his questioner moved some distance from him. Once when Mr. von Osten thought the horse's hearing was being tested, by placing the questioner successively further away, Pfungst was studying the animal's vision. At several meters there was a decline in correct responses, and this pattern increased with greater distance.

It seemed to Pfungst that perhaps the questioner made extremely minute movements, unnoticed by all except the horse. Maybe even the questioner himself was unaware of his signals. Pfungst thus continued to alter systematically the ways in which the horse viewed his questioner, and finally he found the answer. The horse was using extremely minute visual clues.

Law of Parsimony Without realizing it, a typical examiner bent forward ever so slightly from the waist after presenting a question to the horse. Then, when the horse reached the correct number of taps, the examiner bent backward and upward ever so slightly. The horse began tapping at the first involuntary movement and ceased tapping after the second one.

Pfungst readily demonstrated the animal's dependence on these signals. He asked the horse to count to 13 but leaned forward until 20 taps had occurred, which the animal gave without hesitation. He then asked Clever Hans to add 4 and 3 but did not make the upward and backward movement until 14, which of course was Clever Hans' answer.

Similarly, whenever Hans was asked to point to an object or pick it up, he was signaled by the inclination of the questioner's head and upper body. The questioner unintentionally cued him in that direction, just as he signaled the horse to move his head back and forth when the answer was "zero." This finding also explained Hans' alleged capacity to take the other person's viewpoint. When the examiner facing Hans said "left" and the horse looked to his right, he was merely following the questioner's unintentional signal, not the word. And it also accounted for Clever Hans' unusual ability with mathematics. The old schoolteacher was highly adept —not his pupil.

Of special significance, however, is the fact that neither Mr. von Osten nor anyone else, prior to Pfungst's work, was aware of these extremely minute, unintentional signals. It is this subtle, unconscious cueing that separates this case from countless others seemingly similar on the surface.

Many people, including Mr. von Osten, had been aware that gross and intentional body movements can influence an animal's

70

response. Rosa, the famous circus mare in Berlin, was obviously the product of direct instruction, trained to obey her master's signals, which anyone could observe with dogs, pigs, monkeys, and other circus animals. But not even Mr. von Osten himself, who willingly underwent the investigation, was acquainted with the movements he transmitted to Clever Hans, and typically they could not be observed by spectators, even after they were informed of them.

It thus became clear that the original report of the Hans' Commission was misconstrued by practically everyone. This report merely stated that the horse had performed as claimed and that there was no trickery or unintentional influence. As Stumpf said, "The Committee of the 12th of September never in a single word ever maintained the intellectual capacity of the horse."

But a leading scientific journal, through hearsay or its own error, gave a very different impression. It stated that the Commission had unanimously supported "the mental powers of the animal." A newspaper announced that German scholarship had embarrassed itself in a grandiose way. Early readers of such accounts, knowing the subtle abilities of certain animals, concluded that Stumpf had been gullible and the Commission incompetent.

One man composed a lengthy objection to Stumpf's work, citing a London horse, a Tennessee pig, and a local collie, all of which performed through subtle tricks. The dog spelled out answers to questions by picking out letters of the alphabet printed on cards strung in a row. In selecting each letter, the animal always began at A, however, and with good reason; his master, gloves in hand, gave a light but obvious twitch of the gloves as the dog trotted past the correct letter.

Stumpf lamented these distortions in the daily press, which misled the public and tarnished the reputation of his Psychological Institute. Earlier the Commission had ruled out all possibilities of this sort. As he said, thirteen men of varying scientific preparation could hardly have progressed further. To have acted as anything more than a court of honor for Messrs. von Osten and Schillings would have been almost impossible for this diverse group. The intricacies of Pfungst's careful experimental studies were certainly beyond its scope.

71

The key to these signals, of course, was the questioner's desire that the horse perform well. Pfungst called it a high degree of expectancy, and eventually he observed it even in himself. He noted an increasingly unpleasant feeling as the designated tap came nearer and a relaxation when it occurred. This change in tension was manifest as the unconscious signal.

In a series of tests to examine this tension, three conditions were established, apart from the state of complete relaxation. With the lowest degree of tension Pfungst obtained many wrong answers, often by several units too many. With somewhat greater concentration the answer was too large by one or two units. Pfungst then employed the greatest concentration possible, and here the answers were usually correct. The most favorable degree of tension was one in which the concentration was strong and steady but not severe. It is easy to understand why Pfungst acquired headaches in the early course of this work, but after a few days they disappeared and did not recur.

These results explain why Clever Hans at times performed poorly with the number one. It is difficult to relax immediately after beginning to concentrate. The moment of change in tension and jerk of the head occurred too late, with the result that the horse tapped twice or more.

Pfungst's findings also illustrate an important principle in science, the *law of parsimony*, which states that among equally plausible explanations, the simplest is to be preferred. Extraordinary interpretations, such as ESP and psychokinesis, are to be avoided whenever possible. The idea in science is not to complicate things needlessly.

A specific application of this law, called *Morgan's canon*, warns against interpreting any behavior in terms of a higher mental process when it can also be interpreted on the basis of a lower one. According to this principle, Clever Hans should not be credited with thinking if his behavior might result from memory; he should not even be credited with memory if the behavior might be merely a learned reflex or habit, which indeed was the case with the horse of Mr. von Osten.

Testing Clever Hans. *Three men examine Clever Hans and record their data while Mr. von Osten stands at the right. These data were collected after Pfungst's tests.*

How then does one explain the horse's failures? If Hans' success was nothing but habit, what caused his mistakes? Where was the fault here?

Ad Hoc Hypotheses

Pfungst speculated that every error would be a reflection of the questioner, not of the horse. An hypothesis of this sort, developed after the research is underway, is called an *ad hoc hypothesis*. It emerges in the course of the investigation, primarily to account for some specific problem or inconsistency in the findings. According to Pfungst, if success lay with the questioner, so did the mistakes.

Here there were two possibilities, each of which could be readily tested. First, Hans' errors might be the result of errors in his questioner's thinking. In fact, a review of the data showed that sometimes both the horse and his questioner had been corrected by members of the audience. After the person became aware of his error and asked again, Hans answered successfully. These mistakes were clearly of human origin.

Second, there was the possibility that the examiner transmitted the signal poorly or incorrectly. Using the best timing devices of his day, which recorded intervals as brief as one-fifth of a second, Pfungst discovered this to be the case. When the examiner's involuntary upward movement occurred too soon, Hans stopped tapping too early; when it came too late, he gave too many taps. Further inquiries showed that the examiner had become distracted or that his attention had wandered during the task. Again, human frailties seemed chiefly responsible.

These two sources of error played a most important role in the fanfare that surrounded the horse. Pfungst found that a highly excited questioner might have the wrong answer in mind or might lose his concentration, and in very rare instances these errors occurred together, exactly compensating for one another. The befuddled questioner gave the signal for a wrong answer at the wrong time, but it was the right moment for the correct answer, much to the chagrin of the human being. The horse was much applauded by those in the grandstand on these occasions.

Such fortuitous circumstances apparently occurred when Count Otto zu Castell-Rüdenhausen asked the horse the date, thinking it was September seventh. When he gave that signal a bit late, Hans correctly answered "8." Similarly, the flustered Count miscalculated an arithmetic problem, and then gave the signal too late, prompting Hans to continue to 16, which was the correct response. These coincidences, which occurred no more than a half-dozen times, greatly furthered the cause of the celebrated horse.

The inescapable conclusion in all this was that no trickery was involved, that Hans' intelligence had not increased, and that the horse was not telepathic in any way. The only difference between Clever Hans and other horses was that he had received extraordinarily intensive instruction in reacting to a visual signal. Even in Schillings' telepathy tests, the zoologist himself gave the start signal and the person who knew the number indicated when it had been reached.

A forward tilt of the head was the signal to begin, an upward movement was the signal to stop, and a carrot was the reward for tapping between these signals. Without knowing what he was doing, Mr. von Osten had been rewarding the animal for this simple habit, all the while believing he was teaching him to think. Behavior modification had been involved, but not in the manner that Mr. von Osten had expected.

In his final report Oskar Pfungst stated that the animal's capacity to perceive his master's minimal movements, unintentional as they were, was astounding to the highest degree. They went unnoticed even among the sophisticated members of the Hans Commission. Detection was made all the more difficult, Stumpf pointed out, because Messrs. von Osten and Schillings, in particular, moved hither and yon in a most irregular manner while putting the horse through his paces.

It was speculated, however, that Mr. von Osten's headgear played an important role in the animal's early training. This broad-brimmed hat brought to the horse's attention the slightest change in his master's position. Any movement of the man's head was greatly magnified in the arc transcribed at the edge of his large hat.

In response to all this, Mr. von Osten became more emotional than ever, denouncing both the public reaction and the scientific inquiry. "I have witnessed the spiritual and intellectual development of my pupil," he said "and now there come people who want to place restrictions on my work. . . . If we all showed such little intellectual initiative, now-a-days we would still be living in trees."

Pfungst's report absolved him of any attempt to deceive the public, emphasizing that he had cooperated readily in the investigation and that he had stood merely to lose by scientific inquiry. But the unpredictable trainer was also greatly saddened by the result. The remarkable product of his efforts was suddenly exposed as a simple error on everyone's part, and his immense grief was to be expected. He had not taught the horse very much at all, though he had taught humanity a great deal.

CHAPTER SIX

Verification

So Clever Hans was an intelligent horse that proved to be not so intelligent. We could lead this horse to water, but we could not make him think. At least he could not think like a human being and certainly he was unable to read someone's mind. He was an astute observer of certain movements and no more.

But if Mr. von Osten had taught humanity something about horses so had Oskar Pfungst taught us something about the scientific process. As one newspaper reported, there could hardly have been a more striking opportunity for experimental psychology to demonstrate to a broad public the capability of its method.

The crucial factors had been identified by careful manipulation of the conditions of the horse's performance. This manipulation is the essence of the experimental method. It allows the investigator to make statements about what leads to what and thereby to identify cause-and-effect relationships. For this reason the experimental method is generally regarded as the most powerful of all research methods, and Pfungst's work is considered a classic demonstration of its value.

After his work with the horse, Pfungst did something else that illustrates the scientific process. Finished with the horse, he left the courtyard, went indoors, and repeated his work in a new setting, where he could manipulate in an even more precise way the sights, sounds, and smells available to the subject.

Sending Signals   In a word, Pfungst was intent on *verification,* in which a finding is confirmed through further tests. The purpose is to determine the extent to which the earlier result is constant and universal. This process is also called replication, meaning that the research is repeated, presumably with the same basic outcome.

For this purpose he studied in his laboratory at the University of Berlin twenty-five people of different ages, both sexes, and diverse backgrounds. They included children five to six years old and persons of various nationalities, and in leaving Clever Hans behind there was still another advantage. Pfungst avoided horse bites. He had been nipped several times by poor Hans under the discomfort of the huge blinder.

This renewed effort involved two specific questions. First, were the expressive movements of Messrs. von Osten, Schillings, and the others typical of most human beings? Second, could people learn to observe these signals, if indeed they were fairly universal? Affirmative answers to these questions would support and extend his earlier work.

His approach at the outset was much the same as before, except that the volunteers served as questioners and Pfungst played the role of Clever Hans. This procedure allowed him to study various people sending the signals and to observe his own reactions in receiving them.

The questioner, unaware of the purpose, stood just to the right and a bit in front of "Pfungst the horse." As the question was asked, Pfungst watched carefully, prepared to tap out the answer on the table with his right hand, which was more convenient than his foot. In the first tests the questioner selected some number up to ten and concentrated upon it. Without knowing that number, Pfungst began to tap and soon found what he expected: a sudden involuntary movement of the questioner's head, less than one millimeter in distance. When Pfungst ceased tapping, it usually turned out that the correct number had been reached. Altogether twenty-three of twenty-five questioners signaled in this way, and after some early practice Pfungst had no trouble "reading" their minds.

None of the subjects knew Pfungst's purpose and, when asked directly, none noted any particular movement for which Pfungst was searching. Only in rare instances did they report any movements on their part. They were quite puzzled by the proceedings and insisted instead that no cues had been given.

So much for numbers. Persons might give an involuntary signal when a certain value was reached, but what about more complex answers? What signals would be given?

New subjects were asked to concentrate on any one of six possibilities: up, down, left, right, no, or yes. Each subject could think about the concept in any way possible and could even insert a "blank" trial on occasion, thinking of nothing in particular. The latter procedure made the task more difficult for Pfungst who, after a ready signal, attempted to guess the person's thought.

Sometimes he answered by shaking or nodding his head, or by pointing in a certain direction, just as Clever Hans had done, but often he simply said the word he thought the questioner had in mind. Again, statistics told the story. With 12 questioners and 350 trials, his score was 73% correct, and with certain persons he was right almost every time.

In these tests the slight, involuntary head movements were accompanied by involuntary movements of the eyes that showed little variation from one person to another. The four directions—up, down, left, and right—were expressed by coordinated head and eye movements in these directions. Similarly, "yes" and "no" were indicated by nodding and shaking the head, respectively. These movements always took place when the subject began to think about the concept, and they too occurred without the person's awareness. Even after Pfungst disclosed his secret, most persons found it impossible to inhibit them, while the blank trial generally yielded little or no expressive movement.

The quantity zero was tested and similarly detected, and here the eyes were less important. Zero was expressed by a back-and-forth or oval movement of the head in the air, though with practice Pfungst could sometimes discover whether the digit or the word had been conceived, depending on the movements. He noticed other expressive behaviors, such as dilation of the nostrils and twitching of the limbs, but the head and eye movements were most important.

Pfungst then changed his procedure once again, making the task still more difficult by selecting pairs of similar words. He asked each subject to think about "Kürbis," "Kiebitz," "Ibis," or "Irbis," meaning pumpkin, plover, ibis, or panther. Pfungst would indicate which of these four words he thought the subject had in mind by moving his arm to the left or right, forward or backward, respectively. In twenty tests with one female subject he left her completely amazed. He was right fifteen times without arousing in this person any idea that she was actually signaling him through unconscious movements of her eyes and head.

The special significance of these tests was that Pfungst had established an arbitrary connection between the concept and a given unconscious movement, and again the case was conclusive. After succeeding with the young woman, he used different movements with other subjects. He guessed the word "Kiebitz" using an upward movement of his arm in one instance and turning his head to the left in another. Sometimes he said the answer even before the test began. Imagine the subject's surprise when he said, before the ready signal, "You had thought of concentrating your mind upon 'Ibis' but later decided in favor of 'Kiebitz.'" In two series of 35 trials he made no errors at all.

With these results Pfungst had further evidence for the sending of signals. He had identified the natural signals noted in the courtyard, and he had found that certain arbitrary signals could be induced. But could the arbitrary signals be made to replace the more natural ones? What would happen if Pfungst made arm movements up and down to indicate the directions right and left, respectively? Would the subjects use the natural right-left movements, or would they adopt Pfungst's up-down mode of signaling?

Again, the question was readily answered. The subjects began with the natural right-left eye signals, but soon these were accompanied by the arbitrary head movements. The subject indicated "left" with an eye movement in that direction and a downward motion of the head; "right" was signaled by eye movement to the right and an upward motion of the head. Altogether, Pfungst made 32 correct guesses in 40 trials, even when he placed a blindfold on the subject and observed only the induced head movement.

The subjects surely must have decided that Pfungst was a strange man of little whims when next they were told that the signals would be reversed. Pfungst would answer "left" with an upward arm motion and "right" with a downward movement, and after a dozen trials under this new system the former head movements were completely displaced by the new ones, totally the reverse. Finally Pfungst ceased arm movements altogether, indicating his guesses merely by saying them. The expressive head movements continued for a time, then became more tentative, and finally disappeared. It was thus clear that arbitrary signals could be formed, that they could be superimposed on older ones, and that they could be eliminated.

They even occurred with abstract words, not associated with any particular visual image, such as "Form," "Inhalt," "Masse," and "Zahl," meaning form, content, measure, and number. Pfungst indicated these guesses by various movements, and after a while the subject fell into the expected pattern, executing the movements unconsciously. The results with one person, a psychologist trained in experimental introspection, were most impressive. No matter how closely he observed himself, even this professional subject was unable to discover the source of Pfungst's success.

So Pfungst used different people, a new "horse," and special apparatus, all to verify and amplify his explanation of Clever Hans' horsesense. To avoid mere opinions as to whether someone had moved his head, he employed instruments previously used for detecting hand tremors and adapted them for head movements as small as $\frac{1}{10}$ of one millimeter. He also used instruments for detecting breathing and pulse rates.

Recording Apparatus. *This instrument, an early version of the modern polygraph, was connected by wires to various parts of the subject's body. As the drum slowly turned, the pens recorded the subject's breathing, pulse rate, and other movements, while the metronome, at the right, marked the time.*

Some people were found to be inhibited; others showed unusual reaction times; and still others made extraneous movements. Toward the end of his work in the courtyard, Pfungst had noticed that Mr. von Osten's movements were distinct and constant, varying little from one instance to the next, but Mr. Schillings' were less predictable. At times the African adventurer raised his entire upper body and the signals could be noted from behind. He even showed a tendency to nod his head at each tap of the horse, almost keeping time with Clever Hans' answer, and then the last nod, most pronounced, was followed by the typical upward thrust.

In the laboratory, using this special equipment, a rather uniform pattern was discovered for most subjects. Breathing typically remained unchanged; the forward-downward movement of the head was slight; and the backward-upward signal was more pronounced. But there were individual variations in the style and economy of this response.

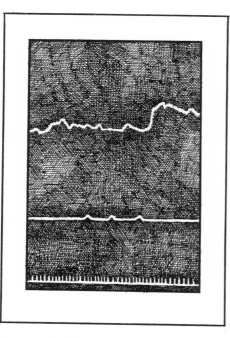

The Signals. *The top line displays the head movements of one subject, Mr. Chaym; the middle line shows Pfungst's tapping; and the bottom line shows the passage of time in one-fifth seconds. Thinking of the number three, Chaym gave a slight downward movement, indicating that Pfungst should begin, and later a more distinct upward movement, which told Pfungst to stop.*

Altogether, these unconscious movements were only half the answer, however, for there was still the question of detection. How was it that only Clever Hans noticed them? Why were they overlooked by the thirteen respected men of the Hans Commission?

Here we are concerned with receiving the signals and a most fundamental consideration: the difference in visual capacity between the horse and human being. This topic falls in the domain of *comparative psychology*, which compares human and animal behavior or the behavior among different animal species. The basic premise is that an understanding of one species or one function, such as vision, may be gained partly through knowledge of other species.

Comparative Psychology

In fact, the difference in vision between the horse and his human spectators played a vital role in making Hans seem clever. It is a thin thread by which hangs the whole celebrated tale and it concerns the location and construction of the eyes.

The eyes in human beings are both at the front of the head and close together. They normally look at the same object at the same time, a condition known as *binocular vision*. Focusing in coordinated fashion from slightly different locations, each eye records a slightly different perspective of a nearby object, seeing somewhat more of one side than the other, but the two eyes have almost the same view of distant objects. This limited capacity to "see around" a nearby object gives the viewer some idea of its depth or distance.

The horse has two eyes in very different locations, one on each side of the head. It rarely moves both eyes together, and they diverge even when it attempts to do so. It relies instead on *monocular vision*, observing each scene with just the eye on that side of the head. For this reason the horse is somewhat handicapped in the perception of distance, but it has an extraordinarily broad *visual field*, which refers to the breadth of vision at one point of fixation. It can see forwards, backwards, and all around at the same time, as

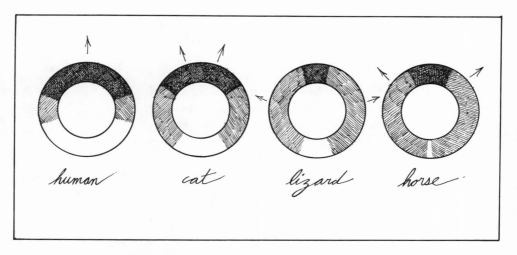

Visual Fields. *The light gray area shows monocular vision; the dark area indicates binocular vision; and the arrows show the direction of normal gaze.*

many an unfortunate soul has discovered by surprising a horse from the rear.

In human beings the visual image is recorded through two types of mechanisms. In the center of the retina the cones provide color vision; toward the periphery, the rods are concerned with black, white, and gray. The rods also possess special sensitivity to movements, which is why you sometimes see something out of the "corner of your eye" that might otherwise be missed. When a stimulus is brought slowly from the periphery toward the center of your visual field, first you can tell that something is moving, and later you can identify its color and shape.

In the serenity of his laboratory, Pfungst noted this characteristic of human vision. It was not essential to look directly at the questioner in order to perceive the signal. When focusing his gaze to one side, so that the image of the questioner's head fell on the periphery of his retina, he was still correct on 90% of the trials.

The horse, with eyes larger than those of any other land mammal, has many more rods than human beings. Since the sequential stimulation of rods underlies visual detection of movement, the horse is usually adept in this regard, a sensitivity certainly important to wild horses escaping predators. But the horse pays a price. With comparatively fewer cones, it does not have our capacity for color vision.

Standing in that Berlin courtyard were people with normal human vision, their eyes darting hither and yon in coordinated fashion, trying to understand what was happening. But humanity, with its own particular view of the world, could not have the horse's perspective.

In the center, amid the babel of sounds and sights, which were nothing but noise and colorless objects for him, stood Clever Hans, particularly on the lookout for the rapid minute movements to which he had become especially responsive. His large and motionless eyes served well in this carnival-like atmosphere, helping satisfy his hunger and whatever pride had been developed in him by the extensive training of Mr. von Osten. Even when approaching some distant cloth, he easily kept the slightest movements of his questioner well within his sight.

Like most of his species, the iris of Clever Hans' eyes was darkly pigmented, barely distinguishable from the pupillary opening. Had he by chance possessed the so-called glass eyes occasionally found in horses, in which the iris is light or colorless, never would he have gained any such position of prominence in the animal kingdom. The dark pupillary opening would have been obvious against the lighter background, indicating the direction of his gaze.

Even the white sclerotic coat surrounding the pupil, more visible in the human eye, is largely hidden by the horse's eyelids, except when the eye is greatly turned. Pfungst once tested Clever Hans in this regard, stepping to the rear of the horse and asking for the animal's attention. To maintain visual contact, Hans turned his eye far backward. Then one could see the white of Hans' eye, and there was no doubt about his line of vision.

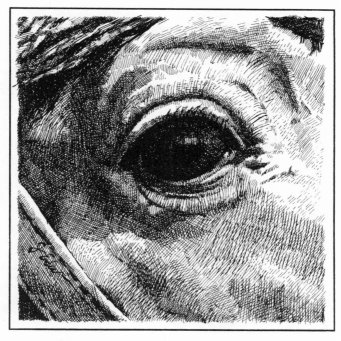

Han's Eyes. *The horse surveyed the entire scene with largely motionless eyes, using different parts of each huge retina.*

In the dog and many other animals the problem could not arise in the first place. The eyes and even the head are almost always turned in the direction of the object being observed. And with this excursion into comparative psychology, the mystery of the horse is further resolved.

But back in the laboratory Pfungst was still unsatisfied. He had *Observing* learned to discern these cues, but could others do so too? With *Signals* training, would this ability prove fairly universal?

The key was being sensitive to an expressive state in someone else, which Pfungst had already examined in himself. From his own experience he knew it was characterized by muscle tensions in the neck and head, gastric sensations, and a rising feeling of unpleasantness until the final tap was reached. In "reading" or detecting these reactions in others, he studied chiefly the neck and head, including the eyes and nose, since the gastric condition and unpleasantness were not directly observable. But all went for nought if the other person failed to concentrate adequately.

So Pfungst and his subjects exchanged roles. He took the part of the questioner and they played the "horse," after he had explained and illustrated the expressive signals.

In one instance, he laid five sheets of colored paper in a row on the floor and stood about seven meters away, concentrating on them. The subjects performed consecutively, attempting to indicate which sheet Pfungst had in mind, and some of these people were notably successful.

In case his interest in the project prompted him to overly zealous movements, Pfungst later had someone else serve as questioner. With a half-dozen persons in this role and two hundred trials, most "horses" showed themselves adept at reading the signals. Only once was the error more than one position to the right or left, and the average number of correct responses for the entire group was 77%, compared with 82% for Pfungst alone. This difference probably was attributable to the fact that Pfungst had received more practice than the others. Practice was essential, together with careful concentration, and these findings provided further proof that Pfungst had discovered a general phenomenon, not some special capacity on his part in one role or another.

In these tests Pfungst also noted *individual differences* among his subjects, meaning that they varied from one another in this capacity. Some learned to detect even the most subtle cues, while others responded only to obvious signs. Such differences are inevitable, and a knowledge of them is essential to establishing optimal environments for human performance. Think of the importance of adjustable seats, hats of different sizes, and prescription glasses.

In fact, one early observer of Clever Hans had detected the signals without any training at all. This man of French background, Max Dessoir, wrote in a calm but skeptical manner that he thought he had seen some involuntary cues—"but by no means in all cases."

Despite Pfungst's earlier expectation, the concept of threshold was not critically involved. The whole unglorious and sometimes unpleasant affair had been largely a matter of not noticing the signals, rather than being unable to detect them. With careful training some subjects became highly adept, and Pfungst stated that the signals were "at the fringe" of consciousness, leaving a more exact distinction to others.

Through these efforts Pfungst verified his earlier findings, making the hypothesis about visual signals more reasonable, rather than less so. The same expressive movements and postures noted in the courtyard had been observed in the laboratory among diverse human subjects, and the same mental attitude he found in himself had been reported by others. All this evidence underlay the success of "Clever Oskar."

Anecdotal Evidence   In closing his case, Pfungst looked briefly beyond the laboratory to the world at large. He turned to *anecdotal evidence*, in which support is sought from incidents in daily life, occurring without research controls and usually reported without full details. Such incidents may strengthen a particular viewpoint or serve as a source of hypotheses, but they are not to be trusted alone as verification.

He cited several cases of dogs adept in detecting subtle signals, including one animal named Kepler. This mongrel could solve highly complex problems, barking instead of tapping his answers, and as a reward he received a piece of cake. His master, an English physicist, said that the pet knew when to cease barking by looking at his owner's face, though no voluntary signals were given.

On August 21, at the very peak of Clever Hans' performances, a newspaper described another dog that could ring a bell any specified number of times. His proud master at first controlled the animal merely by thinking of the final number, but when this procedure failed he concentrated instead on each separate push of the button, with the dog gazing steadily into his face. To make the dog ring five times, his master concentrated five times, and they achieved success up to number nine. At that point the dog became impatient and rang the bell continuously.

Anecdotal evidence for the sensitivity of a well-managed animal is also available in literature, and Pfungst cited the famous horse race in *Anna Karenina*. Here Count Vronski, riding Frou-Frou, is about to pass Machotin, riding a horse called Gladiator. Suddenly Frou-Frou begins to respond to the mere thoughts of her master:

"At the very moment when Vronski thought that it was time to overtake Machotin, Frou-Frou, divining her master's thought, increased her pace considerably and this without any incitement on his part. She began to come nearer to Gladiator from the more favorable, the near side. But Machotin would not give it up. Vronski was just considering that he might get past by making the larger circuit on the off-side, when Frou-Frou was already changing direction and began to pass Gladiator on that side."

The reader is left to speculate on the subtle signals, if any. Such responsiveness, Pfungst emphasized, requires a high degree of sensory keenness and vigilance but not necessarily intelligence. He noted the possibilities for tactual, visual, and muscle sensitivity among horses, deciding that auditory perception probably was of lesser importance than generally supposed. When Clever Hans ran free in the courtyard, he paid no attention whatsoever to Pfungst's calls and shouts, though he came immediately whenever Pfungst beckoned.

In reviewing all this information, Pfungst decided that microscopic movements probably played a role in so-called "mind reading," despite claims to the contrary in the United States. Success in this enterprise, he said, depends partly upon careful scrutiny of these involuntary movements, when it is not based on some sort of trickery in the first place. He stressed sight, sound, and even touch, claiming that the mind reader is guided by changes in muscle tone and temperature while holding the hand of the subject.

With these disclosures about Clever Hans, things went back to normal on Unter den Linden and in the spacious Lustgarten.

Military Parade. *The celebration for Kaiser Wilhelm is passing down Unter den Linden with the University in the background.*

Pedestrians and spectators turned to other issues—the Kaiser, the colonies, and the miners in the Ruhr valley. Most of them promptly forgot the case of the remarkable horse. The whole episode had been only a story in pantomime. The animal was the mirror of the man and nothing more. Speech was superfluous; vision was the answer; Clever Hans was simply following the wishes of poor, unsuspecting Mr. von Osten and the other questioners.

These findings illustrate not only the experimental method and aspects of behaviorism but also the potential for human error in science. Just as Mr. von Osten unconsciously signaled Clever Hans, just as Pfungst's laboratory subjects cued one another, and just as Count Vronski perhaps signaled his mount, in the same way any experimenter may signal any research subject. The research result may be due partly to the investigator's hopes and expectations.

This possibility, that an investigator's personality can play a subtle but important role in research, was first demonstrated experimentally and in detail by Oskar Pfungst in this case of the supposedly intelligent horse. This finding in the long run is the single most significant contribution of the Clever Hans research to psychology and to other sciences.

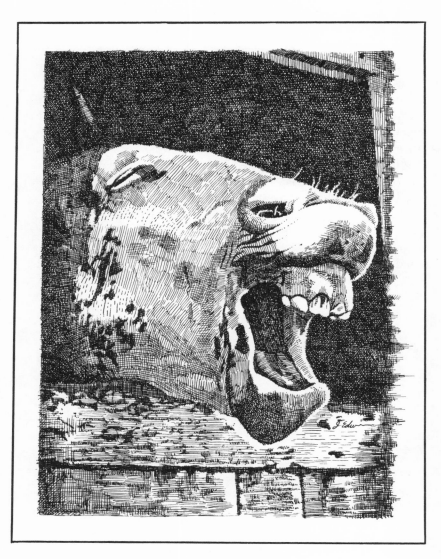

The Horse's Mouth

Part Three
The Strategy

CHAPTER SEVEN

Being Scientific

So Pfungst did some experiments and the problem was solved. He demonstrated to almost everyone's satisfaction that there was no secret Socrates inside that beast of burden. Defenders of any other view, Stumpf said, should not shrink from the task of careful investigation themselves, unless they wished to engage in mere guesswork. For it was through the scientific method that he found an explanation for Clever Hans.

The root meaning of the word *science* is knowing or knowledge; modern science is simply a search for knowledge. Contrary to popular opinion, science is not any particular subject matter, and it is not any special apparatus. Rather, it is a system for finding out about things.

We speak of *the* scientific method, but there is no one method. There are various approaches and procedures for obtaining knowledge in the different fields, though there is only one attitude—that of demanding evidence. When speaking of the scientific method, we are referring to this evidence-demanding attitude.

More precisely, we speak of this search for knowledge as a process, the *scientific process*, for it involves a continuous series of checks and balances available for use in any field, divisible in idealized form into three basic steps or stages. These stages indicate in a general way the fundamental activities in the scientific process.

The system begins with a *hypothesis,* which is a guess, a tentative explanation, an idea to be examined. In the case of Clever Hans, the most common hypotheses concerned N rays, a special animal intelligence, mental telepathy, trickery, and various sensory cues. Each of these ideas gave a possible direction for research, a path to be followed.

Testing a hypothesis, the next step, requires observations of some sort, and these are made in such a way that evidence is gathered for or against the hypothesis. In the courtyard Pfungst examined one idea, then another, and then another. He used the large blinder, then a screen, and then had people whisper to the horse. Each test was aimed at supporting or refuting some hypothesis.

Finally, the result must be verified. Its accuracy or correctness must be determined by repeating the test, as Pfungst did in the laboratory with several groups of subjects. One requirement in most scientific inquiry is that exact repetitions of the original research produce the same result.

These, then, are the basic steps: forming, testing, and verifying hypotheses. They characterize much but not all of science, and they were available to psychologists at the time of the Hans legacy. They proved to be part of the process by which the secrets were revealed, though science does not inevitably follow this idealized form.

More fundamental than these steps or stages are the underlying principles: observing the facts and putting them together. Pfungst, the scientist, and Schillings, the African adventurer, both behaved in this way, observing the horse and thinking about what they had observed.

When thinking is used in this logical way, to obtain further knowledge of something, it is known as *rationalism.* Rationalism is basically the process of reasoning, which down through the ages has played an important role in revealing fundamental truths. Acts of reasoning are involved in all stages of scientific inquiry. The investigator must decide how to find the facts, how to choose among them because they are infinite in number, and how to make some sort of interpretation. Pfungst, in searching for a more and more thorough explanation of the horse, could hardly have done otherwise.

But it is also a basic tenet of modern science that reasoning alone is not enough. Direct experience is essential too; the facts must be observed through the senses, just as they occur. This principle of direct experience, known as *empiricism,* is a cornerstone of scientific research.

The large animal in Berlin was studied in this way, as Pfungst observed how the horse behaved with the blinder, without it, in front of the screen, when the questioner knew the answer, and so forth. In each case he tested the horse and observed the result himself, not relying solely on reason or on the word of some authority.

The importance of direct observation is illustrated in a legendary story about a horse, credited to Sir Francis Bacon. In contrast to Clever Hans, this horse did nothing remarkable. It was simply there to be observed:

"In the year of our Lord 1432, there arose a grievous quarrel among the brethren over the number of teeth in the mouth of a horse. For 13 days the disputation raged without ceasing. All the ancient books and chronicles were fetched out, and wonderful and ponderous erudition, such as was never heard of in this region, was made manifest. At the beginning of the 14th day, a youthful friar of goodly bearing asked his learned superiors for permission to add a word, and straightway, to the wonderment of the disputants, whose deep wisdom he sore vexed, he beseeched them to unbend in a manner coarse and unheard of, and to look in the open mouth of a horse and find the answer to their questionings. At this, their dignity being grievously hurt, they waxed exceedingly wroth; and, joining in a mighty uproar, they flew upon him and smote him hip and thigh and cast him out forthwith. For, said they, surely Satan

hath tempted this bold neophyte to declare unholy and unheard-of ways of finding truth contrary to all the teachings of the fathers. After many days more of grievous strife the dove of peace sat on the assembly and they, as one man, declaring the problem to be an everlasting mystery because of a grievous dearth of historical and theological evidence thereof, so ordered the same writ down.''

The idea of gaining knowledge through direct observation was foreign to the brethren. They relied too much on authority and reason, too little on experience, while the upstart friar sought the answer "straight from the horse's mouth." This approach separated him from his learned superiors, and this tale is perhaps the origin of that expression.

With Clever Hans the facts were needed, and Oskar Pfungst became empirical. He went straight to the horse's math and other abilities to see for himself.

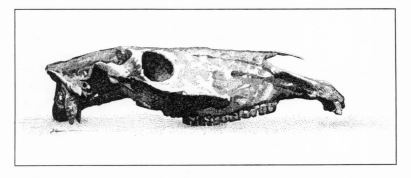

Horse's Teeth. *The skull of Mr. von Osten's first horse is shown from the right side, with the incisors at the front, the normal space, and then several molars.*

Empiricism, Pfungst realized, is essential not only in testing hypotheses but also in forming and verifying them. Like rationalism, it appears at all stages of the scientific process.

Had the brethren followed this advice, they would have found that the colt has only a few milk teeth, which are gradually replaced by twelve permanent incisors. By four or five years the average

98

horse has twenty-four molars, and in the normal adult horse there are thirty-six to forty-four teeth, depending upon the appearance of tusks and canine teeth. The answer to the question, as empiricism shows, is not simple; the number and condition of the horse's teeth vary with age.

But for urging his superiors to unbend at the front end of a horse, the upstart friar was driven from the temple. Criticism was not tolerated in that hall of learning, which is the opposite of the modern scientific attitude.

Years later, when modern science began to detach itself from *Role of* religious, political, and other doctrines, it needed its own guide- *Criticism* lines. It required some standard apart from outside authority and thus came to depend upon *criticism* from within, an openness to opposing viewpoints among members of the field. By careful scrutiny of the methods in any research, and by attempts to repeat the performance, modern scientists determine acceptance within their own community. It is the response and judgment of peers that determines the success of any scientific effort.

When N rays became popular in physics, some scientists accepted this research, as did the public, but as the reports continued and further scrutiny followed, this form of radiation appeared less and less likely. The research did not withstand peer review, and N rays faded from the scene. Science is a community enterprise, a collective search for knowledge, and its results are subject to criticism by those from within as well as those outside the system.

The investigator who studied the boy called Hans was often and severely criticized. Once he was laughed out of a lecture hall, and his research was called "a scientific fairy tale." The man wryly remarked that his colleagues, "whose duties should have been to assist, took pleasure in spitting into the field of operation." His work with the boy remains controversial even today.

It began in Vienna and again a Hans was the central figure, again a horse was involved, the language again was German, and the problem arose near the street. But this question was a practical one, concerning someone's happiness, not a thing of academic or public interest.

Ringstrasse Cafe

Part Four
Hans in Vienna

CHAPTER EIGHT

Another Question

Vienna was the City of Dreams at the turn of this century, famous the world over for pastry, the waltz, operettas, and new architectural designs, as well as cuisine from all parts of the continent. In beautiful parks and public gardens one could enjoy foods with a Latin, Slavic, Germanic, or Magyar past but always with a Viennese flavor. Wiener schnitzel, wiener wurst, and other sausage came to be known as the wiener in honor of Wien, the city of its origin and the German word for Vienna.

Cafes and coffeehouses dominated the city sidewalks, especially on the Ringstrasse, that broad ring of streets encircling the old inner city, each segment with its own historic name. Here one could stroll past magnificent public buildings or spend the morning in the cafes, discussing the latest Strauss or Tchaikowski at the Opera House. Or one could debate the merits of art nouveau, that new artistic style with its flowing lines and decorative floral forms.

At Cafe Landtmann one might even spy the bearded Dr. Sigmund Freud at an adjacent table. His interests in childhood sexuality were scandalous and a lively topic of conversation. No doubt he had just enjoyed a stroll in the Schottenhof district or was fresh from a daily visit to his nearby barber.

On weekends and holidays citizens of this gay city lined the banks and bridges of its chief river, watching young men in striped shirts racing down the Danube in a regatta of four-oared shells. Or they attended the horse races, where gentlemen in top hats and ladies with parasols strolled about the elegant grounds. But most popular of all was a visit to the Prater, that extensive public park offering many amusements, including an exciting ride on the new "Big Wheel."

City Landmark. *Built from the early work of an American named Gale Ferris, this giant wheel gave a spectacular view of Vienna.*

But there was a darker side too. The Hapsburg Empire, nearing the end of its cultural prominence, was experiencing internal problems. Cafe society in Vienna owed some of its popularity to a housing and fuel shortage, for life in public places offered an escape from cold, crowded homes. Viennese culture-makers, such as Mahler in music and Kraus in letters, were villified or suppressed, and

Klimt, the visionary painter, had already withdrawn from a hostile public. Inequities and undercurrents of dissatisfaction had arisen in the long reign of Emperor Franz Josef. Even the mixture of nationalities, to which Vienna owed its cosmopolitan character, had generated political unrest, soon to spread elsewhere.

Into this city in transition came a perplexing problem—a four-year-old boy who almost overnight had become a fearful, withdrawn child. He seemed to have all one might ask, including loving parents and a bountiful home, yet suddenly he began to act in a peculiar manner. His behavior illustrates a common adjustment problem, and it sets the stage for the Hans question that lies ahead, concerning personality. *Concept of Personality*

On the seventh of January in 1908 Hans and his nursemaid left their home to go to the public gardens. When they reached Lower Viaduct Street, in front of their house, the boy became fearful, shrank back, and refused to go further. He asked to be taken inside and then began to cry. The nursemaid could see no threat to the boy or to herself, and he could not say why he was afraid. Finally, she took him back into the house.

Once indoors, his fear diminished, but still he could not say what made him afraid. His usual good cheer returned later, but then around bedtime he became visibly frightened again. He cried and clung to his mother, expressing his desire to coax with her. To "coax" meant to exchange caresses.

The next morning Hans emerged from the house again, intending to go with his mother to the Schönbrunn zoo. His mother's purpose was different; she wanted to discover what was wrong with the child. When they reached the street he became frightened again, shrank back, and once more refused to go. He began to cry, but she persuaded him to continue. Finally, on the way home, he confessed: "I was afraid a horse would bite me."

That evening he cried again and wanted to be coaxed, proclaiming fearfully, "I know I shall have to go for a walk tomorrow." Then he added: "The horse'll come into the room."

Another morning he awoke in tears. His mother consoled him and asked why he was crying. "When I was asleep I thought you were gone and I had no Mummy to coax with," he explained. One evening when he showed this same concern she let him sleep with her, and that comforted him.

Another day he was sent to bed with influenza, after which his fear became even more pervasive. He was afraid throughout the city streets, from Schotten Ring to Stuben Ring, wherever horse-drawn carriages might be found. He would go out only on Sunday, when there was not much traffic. Sometimes he could be induced to make short trips to the country, but once at his destination he refused to leave the garden, fearing an encounter with a horse and buggy.

Schotten Ring Traffic. *Horse-drawn vehicles were used as taxicabs, buses, and moving vans throughout the city.*

This reaction was especially puzzling because Hans had displayed such a calm, self-assured manner in earlier days. He had seemed perfectly content in his surroundings and was almost always on friendly terms with playmates, older people, and even strangers. Apart from this puzzling fear, there were no other disturbances of personality.

The term *personality* refers to an individual's most consistent patterns of behavior. It is the sum total of that person's most characteristic reactions, including interests, abilities, and attitudes. Hans, for example, was a bright, gregarious, independent little boy who had just developed a fear of horses. Whether this fear would become sufficiently prominent and permanent to be considered part of his personality, only time would tell.

In addition to consistency, personality involves uniqueness. Each of us possesses some characteristics in common with other people, some characteristics that are relatively rare and, when the various combinations are considered, a constellation of qualities not found in anyone else. Intelligence, independence, and friendliness are not unusual, but Hans' fear of horses was rather extraordinary. Furthermore, there is a uniqueness in this combination of qualities. The more and less common characteristics unite to form a particular, single whole—a pattern of behaviors marking Hans as different from all other people.

The consistency and uniqueness of any individual becomes apparent largely through interactions with others. There is some regularity in what the person says and does, and some patterns of behavior are different from those of most everyone else. The social context reveals personality, for ways of responding to others are of primary significance.

The concept of personality owes its early meanings to the Latin term *persona,* referring to the masks worn in classical drama. These masks informed the audience about the individual being portrayed. The hero wore a mask suggesting honesty and courage, while the villain's mask implied treachery and deceit. The mask served to reveal, not to conceal, the individual behind it. Today we think of personality as an individual's consistent and unique characteristics, and one wears a mask for the opposite reason—to conceal one's true identity.

It was most unlikely that a horse would come into Hans' room, for he lived in an upstairs apartment in a residential section of the city. It was also improbable that he had ever been injured by a horse, for outdoors he was always accompanied by an adult concerned for his safety. But the boy was afraid nevertheless, and he grew more fearful with each passing week.

Soon he was afraid to talk or think about horses, much less venture outside. He stayed at home, and even then he was constantly on the lookout for them.

A fear with these characteristics is called a *phobic reaction*, for it is both irrational and pervasive. The afflicted individual spends a great deal of time and effort coping with a largely imagined problem. Hence, a phobic reaction, or phobia, may prove to be of considerable significance not only for the growing child but also the adult, depending upon that person's roles in later life.

There is an irony in these irrational fears. Children develop animal phobias, school phobias, fears of the dark, and the like, despite parental efforts to avoid them. On the contrary, they are often unafraid of cars, kidnappers, germs, and poisons, though many parents would prefer otherwise. Altogether, there seems to be a strong maturational factor in childhood fear. The child is not like the giraffe, born with a fear of the lion on first sight, but it certainly is born with a high capacity to learn such fears. The type and intensity of a child's fears depend significantly upon its age and early experience, particularly relationships within the family.

Our understanding of irrational fears is complicated by the lexicon. There are countless words about phobias, stretching from *acrophobia* to *zoophobia*, which mean irrational fear of high places and irrational fear of animals, respectively. The latter term might have been used with Little Hans, except that he was only fearful of horses. There is even *triskaidekophobia*, meaning fear of the number thirteen, but all these words are of little help in understanding the origins of the individual's problem. The problem is merely labeled according to the feared object, which gives no indication of how it arose or what role it plays in the personality.

This proliferation of labels is in violation of *Occam's razor,* a principle named for a philosopher who resisted coining new terms until the necessity has been clearly demonstrated. The reference to a razor suggests that the excess verbiage should be cut away, again stressing parsimony in science. Overly complex terms serve only to confuse the issue.

The problem in Hans' life was also confusing. One Sunday in early spring he mastered his phobia to the point of going to the country with his father. He went along unafraid, but there was not much traffic. "How sensible!" the boy exclaimed, "God's done away with horses now."

Poor Hans. The next week he had to stay indoors because his tonsils had been removed, and then the phobia became very much worse. He ventured onto the balcony but refused a walk of any sort. Upon approaching even the door to the street, he shrank back in fear.

With these events the parents became more and more perplexed, and finally they sought help. The father consulted a friend who could explore the problem more fully, for this man was a *psychiatrist,* specializing in the diagnosis and treatment of mental disorders.

The psychiatrist and *clinical psychologist* are often mistaken for one another because they do much the same work, often together. Both are primarily therapists, concerned with personal maladjustment. But the psychiatrist, with medical training, is qualified in physical treatments and the prescription of drugs, while the clinical psychologist, trained in behavioral science, is better prepared to administer psychological tests and to conduct research.

Both specialists are concerned with deviant behavior, one of the oldest areas of modern psychology. Just as the experimental method emerged from early research in sensory psychology, so the clinical method evolved through early studies in *abnormal psychology,* involving behavioral and emotional disorders. In fact, the psychiatrist's work with the little boy was a landmark in this respect. It was considered so important in the development of abnormal psychology that eventually it became known throughout the field as the case of Little Hans.

Diagnostic *Problems*
The phobic reaction in Little Hans' day was one of several puzzling personality disorders. The puzzle arose because in psychology and medicine it is generally assumed that any change in behavior involves some underlying structural change, but none was found in many cases. Some cases of deafness could be traced to injury to the ear itself, and sometimes neurologists traced the disability to a defect in the nervous system. But in other cases there was no discernible physical basis of any sort.

In these instances, when no other defect could be found, a neurological disorder was simply suspected, and it gave its name to such problems. They came to be called *neurosis,* meaning that the nervous system was involved in some unknown way. The term was a vague one, referring to a diverse collection of disorders of uncertain origin, and it developed partly by default.

It solved the puzzle for a time, however. There *was* something wrong with these people, and the concept of neurosis gave the illness some dignity. The term came to be widely used, as in the case of Little Hans, though when it did not refer to some obvious neural defect, it was an admission of ignorance.

The treatment of such problems remained at an impasse for years, but in the late nineteenth century Jean Martin Charcot, a leading French neurologist, developed a broad reputation for his work at the clinic in La Salpetriere. In this poorhouse with thousands of old women as inmates he set himself the task of bringing some order out of this "pandemonium of infirmities."

Charcot began by identifying similarities and differences between the various disorders, and he investigated brain and spinal cord functions. But it was as a teacher that he excelled. For two decades he gathered about him hundreds of students each week to discuss a particular ailment. He commenced these Tuesday Lectures by introducing one or more patients afflicted with a certain problem and providing comparisons with earlier cases. After the patient was dismissed, he described the probable defect in the nervous system,

110

using drawings flashed on a screen by the "magic lantern," a fore-runner of today's slide projector.

In classifying the symptoms of neurosis, Charcot gradually set a new direction in the field. He noted that certain disorders not caused by the organ itself also did not correspond to what was known about the nervous system. They conformed instead to the patient's usually incorrect ideas about human anatomy. Charcot was struck by the possibility that one's mental life might be a factor in these disorders, and he began to promote this view.

If Charcot's lecture were dramatic, his efforts at treatment were doubly so. He reasoned that since a hypnotized person showed certain similarities to the neurotic individual—muscular rigidity, insensitivity to pain, memory loss, and so forth—hypnosis might be helpful in removing neurotic symptoms. He used this method for a time and his reputation gave it some credibility, though it yielded only marginal results. Much more important was the new perspective, the idea of psychological factors in neurotic disorders.

A generation later the parents of the little boy in Vienna reasoned in similar fashion. No physical cause could be discovered for his strange behavior, but they were determined that he should not be bullied or laughed at for this difficulty. That would only make matters worse. They took pains instead to adopt the child's perspective and to try to understand the origin of this difficulty.

The father had already described the problem to his friend. The *Search for* boy was afraid of going into the street and sometimes even out of *Causes* the house. He was often depressed, particularly in the evening, and then he cried and wanted to coax with his mother.

The psychiatrist suggested that the father continue these observations and point out to Hans that his fear of horses was nonsense. He should also take walks with the boy and discuss his personal problems, even bringing up some facts about sex. But the latter information was to be imparted only when requested.

The father encouraged his son to go walking, especially after the tonsilectomy. "Do you know what?" he asked. "This will get better if you go for walks. It's been so bad now because you were ill."

On one occasion the father explained that horses generally do not bite. Little Hans replied, "But white horses bite. There is a white horse at Gmunden that bites. If you hold your finger to it, it bites." The father wondered why his son said "finger" rather than "hand," and later he made a note about it.

Hans then told a story about a child named Lizzie who lived in Gmunden, a summer resort near the Austrian lakes. He explained: "When Lizzie had to go away, there was a cart with a white horse in front of her house, to take her luggage to the station. Her father was standing near the horse, and . . . he said to Lizzie: 'Don't put your finger to the white horse or it'll bite you.'" This incident made a strong impression on Little Hans, and he repeated the man's admonition several times.

But the warning about horse bites did not explain Hans' fear. It had occurred two summers earlier, after which the boy had reminisced pleasantly about those experiences. He showed no fear whatsoever until its abrupt onset eighteen months later.

During these holidays Hans was often without his father, who stayed at work in Vienna. The boy remained at home with his mother, and when he fell into moods of fearfulness or tenderness, she consoled him. Sometimes she let him sleep in her bed, which seemed to allay his concern.

Shortly after Hans' description of the Gmunden incident a new maid arrived in the household. Hans got on well with her, riding around on her back, calling her a horse, and yelling "Giddyap, giddyap." But the idea of taking just a few steps into the street still prompted great resistance.

One afternoon the father decided to query Little Hans in further detail: "Which horses are you actually most afraid of?"

"All of them," Hans replied.

But the father insisted, "That's not true."

Whereupon Hans revealed, "I'm most afraid of horses with a thing on their mouths."

"What do you mean? The piece of iron they have in their mouths?"

"No," Hans replied. "They have something black on their mouths." And with this remark he put his hand over his own mouth.

"Have they all got it?"

"No, only a few of them."

The father, still puzzled, asked again: "What is it that they've got on their mouth?"

"A black thing," the boy responded, which made the father decide that Hans was referring to the thick harness or muzzle that some horses wore over their noses.

Was there anything else that prompted this fear? The boy then revealed that he was upset by what horses wore next to their eyes. He seemed to be thinking of the blinders on the harnesses of work and show horses.

The father, following his friend's advice, was searching for *the* cause of the fear, and occasionally a single, predominating factor exists. More commonly, it is difficult to identify just one factor with any certainty. Instead there seem to be several contributing causes, each influential in some way.

These factors, if they occur before the disorder, are known as *predisposing factors,* for they create fertile grounds for some problem. The predisposing factors in Little Hans' case might have included physical illness, conflict between the parents, the frightening story about horse bites, and perhaps the birth of a sibling. A child in these circumstances may be insecure already and perhaps predisposed to some sort of maladjustment. There is a tendency towards a disorder.

When some additional factor arises, it may become "the straw that broke the camel's back." If so, it is called the *precipitating factor,* meaning that it is the most obvious causative event immediately preceding the disorder. As a rule, however, an event does not become a precipitating factor without several predisposing conditions. In fact, the precipitating event might have been merely a predisposing condition had it occurred earlier, and some predisposing factor might have been the precipitating event had it occurred

Horses in Harness. *The product of painstaking breeding and training, these Viennese horses pulled heavy loads in a grand style.*

later. Little Hans' experience with the warning in Gmunden might have been a predisposing or precipitating factor, depending upon when the other events occurred.

The father, at one point, decided that Hans' confusion about sex was contributing to the problem, and he explained to his son in a suitable way something of the difference between males and females. Afterward Hans remained in high spirits all day, even playing outside. Toward evening he became depressed, however, and once more complained about horses.

Later Hans said that he was afraid of horses pulling loads of furniture, and the father began questioning him again.

"Why?" he asked.

"I think when furniture-horses are dragging a heavy van, they'll fall down," the boy explained.

"So you're not afraid with a small cart?"

"No. I'm not afraid with a small cart or with a post-office van. I'm most afraid too when a bus comes along."

"Why? Because it's so big?"

"No. Because once a bus horse fell down."

"When?"

"Once when I went out with Mummy in spite of my 'nonsense'; when I bought the waistcoat."

"What did you think of when the horse fell down?"

"Now it'll always be like this. All horses in buses'll fall down."

A moment later the boy revealed, "That was when I got the nonsense."

The father protested: "But the nonsense was that you thought a horse would bite you. And now you say you were afraid a horse would fall down."

"Fall down and bite."

"Why did it give you such a fright?"

"Because the horse went like this with its feet. It gave me a fright because it made a row with its feet." The boy then gave a demonstration, falling to the ground and kicking in the air.

"Was the horse dead when it fell down?"

"Yes!"

"How do you know that?"

"Because I saw it." And then he laughed. "No, it wasn't a bit dead."

"Perhaps you thought it was dead?"

"No. Certainly not. I only said it as a joke."

The boy had his playful side and seemed tired with the questions. He ran off, and the father concluded that both of them were still confused.

CHAPTER NINE

Background

"My dear Professor," wrote the father to his friend one day, "I am sending you a little more about Hans—but this time, I am sorry to say, material for a case history. As you will see," he continued, Hans' "nervous disorder . . . has made my wife and me most uneasy, because we have not been able to find any means of dissipating it."

His letters continued in weekly installments for several months, each containing extensive notes on the boy's behavior. The father is therefore the person to whom we are indebted for the facts about Little Hans, and his friend eventually provided an interpretation of the case, which added greatly to its value.

Once when his father was making these observations Little Hans asked, "Why are you writing that down?" The man explained that he would send the information to the Professor, who would take away Hans' "nonsense." They both adopted this term after it had been used by the Professor.

"Oho!" the boy responded, not wanting to be singled out for

special scrutiny. "So you've written down as well that Mummy took off her chemise, and you'll give that to the Professor too."

"Yes," the man answered, for the boy's remark immediately gave that event further significance. But he continued with his questions and observations, convinced that all sorts of childhood events can be significant for the growing personality.

This view has a long history. Philosophers and poets have written for centuries that "the child is father of the man" and "as the twig is bent, the tree's inclined." In the last hundred years modern psychologists have been studying these details more precisely, attempting to discover what kinds of bending lead to what kinds of inclination. Charles Darwin even played a role, publishing an infant biography of one of his children, the first of its kind. Such observations form the basic data of child psychology.

For any careful observer the most obvious childhood changes occur in physical development, as the muscles and sense organs continue to grow. These structures enable the child to respond to its world. In attempting to understand any child, including one with some behavior disorder, physical development is the usual point of departure.

Physical Development   Some of these physical changes are fairly predictable, especially those due to a biological unfolding called maturation. In such cases we speak of *norms* or normative data, meaning a set of standards for what is normal or expected. In many countries in the Western world, for example, norms have been determined for the *development of locomotion*, by which a child becomes capable of moving from place to place. Every baby begins postnatal life lying down, unable to go anywhere by itself, but it predictably transcends this state.

The infant first raises its chin and later its chest as well. At approximately four months it can sit with help, and around eight months it stands with assistance. Its first form of locomotion occurs at nine months, when it begins to crawl on all fours, an event that keeps the parents moving as well. Just before the end of the first year the average child is able to walk with support, and around fourteen months it first stands alone. Walking, the normal form of

Early Childhood. *Photographs of Little Hans are unavailable, but this picture of a Viennese boy of approximately the same background and era illustrates his development at about four years of age.*

adult locomotion, occurs about one month later, although refinements in running, climbing, skipping, and jumping occur for many years.

To appreciate the significance of these data, three points should be kept in mind. First, norms represent averages for groups of people. Individuals develop at different rates, resulting in wide variations even among normal children. Consequently, and this is the second point, it is the sequence and not the age that is most predictable. Almost any baby will sit before it can crawl, for example, and walk with support before it can stand alone. Third, certain variations in style or mode of response are not unusual. Some normal children never crawl in the true sense of the word. They develop a "hitching motion" instead, moving about on all fours in asymmetrical hops or squiggles, somewhat like a rabbit or inch worm. In short, norms are only guidelines. Individual variations are common and inevitable, depending upon maturational factors and opportunities for learning.

Four-year-old Hans showed no unusual deviations in this respect. He was still improving in strength and coordination, but the basic locomotion skills had been acquired. In the summer he ran and played tag, and in winter he was learning to skate. Some psychologists would argue that herein lies a fundamental purpose of such games. They give children the opportunity to acquire physical, social, and intellectual skills important for later life.

Another illustration of norms for physical development is found in the *prehensile grasp*, which involves taking hold of something with the thumb and forefinger in opposition. Monkeys have a digit comparable to the human thumb, but they cannot use it against the forefinger. The prehensile grasp is virtually unique to human beings.

The human infant shows only aimless movements of the limbs in the first months of life. The hands and feet, fingers and toes, are essentially useless. Later it reaches to grasp things sporadically, and by six months it does so consistently, though the hands are used largely for "palming things," without employing the fingers separately. The thumb is finally opposed to the fingers collectively at around nine months, but only after the first year can the average child oppose just the thumb and forefinger, as in picking up a cookie crumb.

Little Hans had gained far more dexterity. He drew pictures of animals, turned the pages of a book, and dressed and undressed his dolls, which are customary behaviors for a four-year-old child.

Extreme variations of course can be significant. Especially because physical and mental abilities are highly correlated in the early years, marked retardation in physical development sometimes suggests mental deficiency as well. But no significant variations were noted for Little Hans. The boy had been a healthy, active baby whose birth and early growth proceeded according to expectations. He had experienced the usual childhood illnesses, including influenza and tonsilitis, but altogether there was little that marked him as exceptional.

Similarly, Little Hans showed normal *social development*, which is the growing capacity to interact successfully with others. The growing child learns the customs and rules for associating with other people at the dinner table, at school, in games, and elsewhere. These standards of social behavior are transmitted by parents, playmates, teachers, and even strangers in that society. *Social Development*

The child begins to learn social behavior in the first days of life, and by one or two months it will smile at almost anyone. This unselective response prompts some proud parents to believe that theirs is a highly social infant, but the response soon disappears. Around four months only persons who have had extensive contact with the infant elicit this reaction, and the child gestures to be picked up or handled by them.

Gradually, the child learns to ask for further personal attention and to help itself. Not until three years of age, however, does it show obvious signs of cooperation, as at a tea party or in some imagined circumstance, and at just this age a most significant event took place in Hans' social life. His father had predicted it several times, and it demanded a good deal of sharing.

The boy awoke at seven o'clock one October morning and on hearing his mother groaning in the next room he asked, "Why's mama coughing?" After a moment's reflection he answered his own

121

question: "The stork's coming today for certain." These unusual sounds, he decided, were somehow connected to the stork's arrival, but on entering the front hall he saw a physician's bag. "What's that?" he asked. Then he declared more emphatically, "The stork's coming today."

After the birth he was called into the bedroom, whereupon he saw the basins filled with water and blood. Looking about the room, he pointed to the bedpan and exclaimed with surprise, "But blood doesn't come out of *my* widdler."

The baby immediately became an entity in the family and Hans, in customary manner, competed with the young intruder for adult attention. When someone adored Hannah, he observed, "But she hasn't got any teeth yet." On seeing her in the bath, he remarked that she was still quite small, and when he became ill he murmured in a feverish moment: "But I don't want a little sister!"

Little Hans' reaction was not unusual. In fact, competition among brothers and sisters is so widespread that it is known as *sibling rivalry*. Children in the same family compete in various ways, direct and indirect, for the parents' love and approval. Little Hans' younger sister required considerable parental attention, and he did what he could to regain his former position of prominence in the family.

These struggles among siblings are particularly significant because they constitute the child's first sustained encounter with peers. The outcomes of these early interactions presumably influence the child's tendency toward dominance, submission, cooperation, and other ways of responding in later life. The social context, as indicated earlier, is a most significant factor in the formation of personality.

Some months later Hans had surmounted much of his jealousy or at least had converted it to a claim of superiority. Whenever he encountered other little girls, he had a very favorable reaction, one in advance of his years. He was overjoyed when a girl came to visit him in the afternoon and found it hard to sleep before the event. In restaurants, to the amusement of other diners, he flirted with little girls whom he had not met previously and at home, especially after moving to a different apartment, he sat on the balcony steps for hours hoping to observe the girl next door.

Through such relationships, including the sharing of parental attention, the normal child develops the social skills and attitudes that enable him to live with others with satisfaction. He gradually learns accepted social practices.

In the winter after Hannah's birth Little Hans went to the skating rink and met two girls about ten years old, daughters of a friend of his father. He immediately sat down and gazed at these older girls in admiration, though his interest apparently went unnoticed on their part. For days afterward he spoke of "my little girls" and kept asking: "When am I going to the rink again to see my little girls?"

At times he treated girls in an aggressive, almost arrogant manner, hugging and kissing them vigorously, sometimes without warning. When his friend Berta once emerged from another room, he flung his arms around her and said with great feeling: "Berta, you *are* a dear!"

He also showed an interest in boys, especially Franzl and Fritzl, with whom he played all day long at Gmunden. When asked which of the girls he liked best, he even named one of these male friends: "Fritzl."

Gmunden in several respects was ideal for Hans' social development. It afforded a variety of experiences with other children, and he had no such playmates at home in the city. He enjoyed himself with them and, except for Hannah, there was little to suggest that his present difficulties could be attributed to some defect in social development.

At the time of Hannah's birth, however, Hans acquired a lively interest in his reproductive organs. He asked, "Mamma, have you got a widdler, too?"

"Of course, Why?"

"I was only just thinking."

Later, on seeing a cow being milked, he observed "Oh, look! There's milk coming out of his widdler!" And one time at the zoo he joyfully exclaimed, "I saw the lion's widdler." Children's interest in animals perhaps is augmented by the openness with which animals display their organs, satisfying some sexual curiosity in inquisitive children.

Hans' interest extended beyond simply looking. He drew pictures of animals with sex organs, and he played private games in the bathroom. These activities prompted the physician's suggestion that he receive more information about sex.

Following Hans' first fearful reaction in the street, his mother asked, "Do you put your hand to your widdler?" The boy replied, "Yes. Every evening when I'm in bed." He was warned against this practice before his afternoon nap, and when he awoke he was again questioned. A little man of integrity, he admitted that he had broken the prohibition "for a short while."

Prohibitions vary from culture to culture, but learning restraints is part of one's social development. Every society places certain restrictions on behavior, and a child is regarded as socialized to the extent that he has learned what is and is not permissible.

One time when his mother found him in this activity she said, "If you do that, I shall send for Dr. A. to cut off your widdler and then what'll you widdle with?" Dr. A. was not the physician consulted earlier.

Hans replied, "With my bottom."

His mother said ". . . it's not proper."

Little Hans answered, laughing, "But it's great fun."

According to his father, if he refrained from such behavior, his problem might disappear. "You know," the man said, "if you don't put your hand to your widdler any more, this nonsense of yours'll soon get better." The father was thinking that masturbation caused Hans' anxiety, which was then transferred onto the horse.

The boy protested, "But I don't . . ."

"Well, to prevent your wanting to this evening you are going to have a sack to sleep in."

To which the boy replied, "Oh, if I have a sack to sleep in my nonsense'll have gone tomorrow."

But it was not. The fear continued, as did the boy's interests. Down at the station Hans saw someone letting water out of an engine and exclaimed, "Oh, look, the engine's widdling. Where's it got its widdler?" Thoughtfully, he added: "A dog and a horse have widdlers, a chair and a table haven't."

Cognitive   Through such experiences Hans was gaining knowledge steadily.
Development Psychologists call this change *cognitive development*, referring to

124

children's growth in understanding themselves and their environment. Cognitive development is largely mental or intellectual development; it concerns thinking, rather than attitudes, values, and emotional reactions. It involves ways in which children perceive the world, store this information in memory, and then use the stored and newly acquired information in daily life.

Having reached four years of age, Hans was trying to understand certain concepts that adults use more or less automatically in thinking about the world. They speak of life and death, fair and foul, probable and improbable, sometimes without realizing just how difficult these concepts are for the child. Young people struggle with them for a long time before adult mastery is attained.

A *concept* is some general idea or unit of knowledge derived from experience. We have concrete concepts, such as cart, zoo, and tonsilectomy, and also more abstract ones, such as justice, truth, and beauty, which are harder to master. Little Hans in this instance was struggling with the abstract concept of life. He had observed that the dog and horse have something in common; they each possess a sex organ, and in this way they differ from a table and chair. Observing these similarities and differences between living and inanimate things, the boy was beginning to understand what is meant by life.

The process of concept formation involves two steps. It begins with *abstracting,* in which some common feature is observed among several things, such as horses and dogs. Regardless of how they differ in other ways, they have at least this one characteristic in common, which Hans called a widdler. Then the child engages in the second phase, *generalizing,* which means applying this knowledge in new situations. The child observes that the table and chair lack this characteristic; they must be in another category. The steam engine is a doubtful case. It seems to have a widdler of some sort, but otherwise it is quite different from animals.

In this way, abstracting and generalizing, correctly and incorrectly, the child approaches an adult understanding of the concept of life. He notes that movement is a characteristic of being alive and ascribes life to a watch when it runs and to snow when it falls. Later, after age six or more, the child insists on autonomous movement. Altogether, an adequate concept of life takes years to be formed, and even adults are in disagreement about some aspects, such as abortion, euthanasia, and life-support systems.

Walking with his father, Hans once knocked on the pavement with his stick, exclaiming: "I say, is there a man underneath?—someone buried?—or is that only in the cemetery?" He was occupied not only with the question of life but also the riddle of death.

Cognitive development is aided by the insatiable curiosity of children. Persistent little scientists, they learn about themselves and the world around them through constant interaction with the environment. Given a new toy, the child grasps it, squeezes it, bangs it, drops it, shakes it, and so forth, all to understand the effect on the object, on himself, and on his caretaker. The resulting knowledge has been called "growledge" by some cognitive psychologists, emphasizing that the child grows his or her knowledge through such trial-and-error operations.

The most influential spokesman for cognitive psychology has been Jean Piaget, whose career began not long after the onset of Hans' fear. Assisting Alfred Binet in the development of intelligence tests, Piaget soon tired of this task but not of the children, and he went home to Geneva to study them in his own way. Much later, after years of unacknowledged labor, he described cognitive development as a series of successive stages from birth through adolescence. There is a steady unfolding of the child's intellect, he emphasized, much as locomotion, prehension, and other physical skills unfold, and the critical factors in this process are maturation and experience.

Little Hans, not yet five years old, was still in the early cognitive stages. He was increasing his language ability, a task that is characteristic of his age, and then using language in further explorations. Constantly asking the why of this and that, he was carrying out his own research on conception and birth, to which the fable of the stork apparently added more mystery.

The boy had heard his mother groaning and remarked about it. He knew she was confined to bed and exclaimed about the blood in the pans. With no sign of the bird and a doctor's bag in the hall, this thoughtful little boy was skeptical about what he had been told.

He also struggled to understand God, his own father, and countless other matters, some of which might relate to his fear. He even tried to understand his dreams, which on occasion he described in animated fashion.

One night he awoke and came to his parents' bedroom, as he had done on several occasions. They inquired whether he was afraid. "No," he answered, "I'll tell you tomorrow." Then he promptly fell asleep in their bed, from which he was carried into his own room. *Children's Dreams*

The next day Hans reported this dream, which the father wrote down immediately in shorthand: "In the night there was a big giraffe in the room and a crumpled one; the big one called out because I took the crumpled one away from it. Then it stopped calling out; and then I sat down on the crumpled one."

The father was puzzled and tried to find out more. "What?" he asked. "A crumpled giraffe?"

"Yes," replied Hans, fetching a paper and crumpling it up. "It was crumpled like that."

"And you sat down on top of the crumpled giraffe? How?"

The boy gave a demonstration, sitting down himself.

"What can it mean: a crumpled giraffe? You know you can't squash a giraffe like a piece of paper."

"Of course I know. I just thought it. Of course there aren't any really and truly. The crumpled one was all lying on the floor, and I took it away—and took hold of it with my hands."

The father continued: "Where was the big one in the meantime?"

"The big one just stood farther off."

"What did you do with the crumpled one?"

"I held it in my hand for a bit, 'til the big one had stopped calling out. And when the big one had stopped calling out, I sat down on top of it."

"Why did the big one call out?"

"Because I'd taken away the little one from it," the boy responded.

127

The father then asked why Hans had come into their bedroom that night, and the boy's answer was vague. Suddenly the man changed his approach.

"Just tell me quickly what you're thinking of," he insisted.

"Of raspberry syrup," Hans replied in a joking manner.

"What else?"

"A gun for shooting people dead with," he added.

The perplexed father guessed that the raspberry syrup had something to do with Hans' constipation, for which the boy was fed this remedy. But he was puzzled by the dream.

Limitations in children's language make it especially difficult to learn about their dreams. They may be reporting daytime fantasies or actual events in a distorted manner. As a result, it is difficult to know at what age children even begin to dream, though we can observe certain bodily functions presumed to accompany the dreaming state.

After the slow, uncoordinated, drifting movements of the eyes when falling asleep, a person's gaze, even beneath closed lids, seems to shift rapidly in many directions, as though observing some highly stimulating scene. When adults are awakened during these coordinated *rapid eye movements*, they usually report that they have been dreaming. Persons awakened in the absence of such movements are less likely to report dreams. Thus, these indications of the dreaming state have been useful in research.

Little Hans' dream about the giraffes was not unusual, especially for one of his age. Children capable of language commonly dream about animals, and people of all ages dream about recent events as well as earlier ones brought to consciousness by recent experiences. Little Hans had just visited the Schönbrunn zoo, and he had a picture of a giraffe hanging over his head.

Once he dreamed about a holiday at Schönbrunn. He said to his fahter: "I was with you at Schönbrunn where the sheep are; and then we crawled through under the ropes, and then we told the policeman at the end of the garden, and he grabbed hold of us."

The father remembered that they had wanted to visit the sheep the previous day and found the space shut off by a rope. Hans said that he could quite easily slip under the barrier, but his father

Setting for a Dream. *In the Schönbrunn area, outside the center of the city, there were several parks, a zoo, and a royal palace.*

replied that a policeman would come along and take him away. The boy apparently was convinced, for earlier the father had explained that policemen arrested naughty children.

At other times Hans dreamed about being naked on a street corner and about his family. Once he dreamed he was being bathed by his mother and the plumber came into the bathroom and unscrewed the tub because it needed to be repaired. Then the man took a large borer and stuck it in Hans' stomach. Later his father asked about the details of this dream also.

The man walked and talked with the boy regularly, trying to extend their Sunday excursions along Lower Viaduct Street, but

Little Hans could not be induced to go far from his house. One time the father proposed that they go to the Hauptzollamt Station, thence to the zoo, and then to the country. The boy finally agreed, feeling very nervous. Following a piece of advice from his mother, he hurriedly looked away whenever a horse came into his vicinity, which was quite often.

Despite these efforts, the fear continued without significant change. The father was still very much at sea about the whole business, lost in the horse latitudes perhaps, or at least making no significant progress. Finally, he decided that Hans himself should visit his friend. He would take the boy to his friend's consulting room at 19 Berggasse, a street not far away, on a hillside in Vienna.

CHAPTER TEN

Forming Hypotheses

As is so often the case, the lower part of the hill was occupied by the poorer inhabitants of the city. At the top there lived the more affluent and somewhere in between, at number 19, was the office and residence of the friend. According to the name-plate at the doorway, two Sigmunds worked at this address, Kornmehl the butcher and Freud the psychiatrist, both engaged in some sort of dissection. Sigmund Freud used his consulting rooms here for forty-seven years.

Earlier, Little Hans' mother had taken her problems to Dr. Freud, and her bespectacled, mustached husband was one of Freud's early followers. It was natural that they should consult Freud again.

Freud was no ordinary practitioner, however, and freely admitted that he had no interest in playing "the doctor game" or even in helping humanity. The grand concern of his life was "an over-powering need to understand something of the riddles of the world in which we live." It was in this context that he became interested in Little Hans.

Completing medical training in 1881 but with no desire for medical practice, the young physician soon won a travel fellowship for $200, perhaps the most important award in his life, and he went to France to study neurology. Upon arrival he was an experimentalist, more interested in brain anatomy than clinical treatments, but gradually his emphasis shifted. The French laboratory did not suit his taste; he was in need of a steady income; and he had encountered some challenging nervous diseases in clinical neurology. Most important, he was coming under the influence of an exciting teacher, Jean Martin Charcot, then at the height of his career at La Salpetriere.

The French physician was using hypnosis in therapy, stressing that certain neurotic disorders might not necessarily have their origin in some defect of the nervous system. One's memory or mental condition might play the same role as a structural defect in causing the abnormality.

The patient in *hypnotherapy* was placed in a sleep-like trance, characterized by heightened suggestibility, and told that the problem would be cured or improved upon emerging from the trance. When Charcot was treating an individual who was inexplicably blind, for example, he placed the person in a hypnotic state, then discussed the problem, and finally said something like: "Now you will be able to see. When you emerge from this trance, you will see." And often the patient could see.

Freud was fascinated, and once back in Vienna he turned more completely from the laboratory to clinical practice. But he was cautious at first and began with the current techniques, including electrotherapy, medication, and massage. He also tried hydrotherapy, in which the patient is soothed by a bath of constantly flowing water. But when all these procedures seemed of little value, he was prompted to reconsider Charcot's work.

For a while he experienced some success with this little-known method. Persons sometimes were cured with hypnosis, and sometimes symptoms could be made to appear in otherwise normal individuals. But there were problems here, too. Treated patients often returned with the same or new symptoms, and some patients could not be hypnotized. Still another difficulty was that the new technique was not acceptable in Viennese medical circles. Unproven and undignified, it brought considerable scorn upon the occupant at 19 Berggasse. It was the beginning of a notoriety that was to spread from the local to the national and finally to the international scene.

19 Berggasse. *The name of the roadway, sometimes written as two words, means Hill Street, referring to the chief feature of the terrain.*

Most important to Freud was the fact that he learned nothing new and thus remained frustrated in the great goal of his life. The secret to the patient's problem remained. The chief value in hypnosis, he concluded, was its occasional success in bringing into the patient's awareness presumably forgotten events. Freud in these years worked essentially alone, except for occasional collaboration with a colleague who brought to his attention a young woman named Anna O. She had several neurotic problems, and after the usual hypnotic procedure she discussed them with her therapist. Gradually, she found that this "chimney sweeping," as she called it, made her feel better. Talking about earlier emotional experiences, especially reliving them in unrestrained fashion, apparently was helpful, and Freud decided to use the "talking cure" himself.

To aid in recall, Freud questioned the patient directly as she lay on a couch. He laid his hands on her forehead, to urge all the more, and when she could not remember seemingly important events, he pressed on her brow. But there were difficulties here too. Freud could lay his hands on the patient but not on her problem; some patients forgot the very events that seemed most important.

Again Freud changed his method, and again a patient helped him find his way. Elisabeth von R. scolded him for interrupting. She said that he was too direct and asked questions at the wrong time. This revelation moved Freud to still another stage in the development of his method, essentially the final one.

Here there was no direct urging and no laying on of hands. The person was simply told to lie quietly on a couch and say whatever came to mind, however irrational, irrelevant, or irreverent it might seem. She was to let her mind wander in an unrestrained manner, a procedure called *free association* because the sequence of ideas is determined by the spontaneous flow of thoughts in the individual's psychic life.

Freud sat and mostly listened, becoming more restrained in his technique. He interrupted only occasionally with some small query, encouragement, or comment, and in the course of these associations his patients sometimes brought up their dreams. Freud encouraged the discussion of dreams in this same way.

As the techniques of free association and dream analysis grew, Freud's work came to be widely recognized, and he acquired more and more patients, including the mother of Little Hans. Later, partly in recognition of Charcot's Tuesday Lectures, he formed the Psychological Wednesday Society, a group of interested colleagues who met each week to refine the new method of therapy and to promulgate the new theory of personality, both of which were called psychoanalysis.

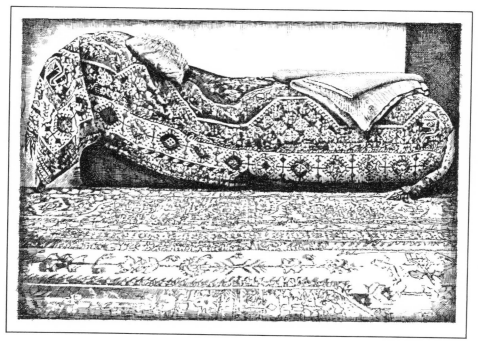

The Couch. *Freud sat in a stuffed chair at the head of the couch, out of the patient's field of vision. The pillow enabled the person to remain almost sitting, if that seemed desirable, and the blanket could be pulled up for warmth.*

As each patient strove to bring to light that which was hidden, *Seduction* Freud began to sense the possible significance of sexual factors in *Hypothesis* human adjustment. At one point he thought neurosis might relate to sexual abstinence in otherwise healthy men or to some interruption in the sexual life of women. But especially as reports of early memories continued, it seemed there was some disturbance in childhood. Freud thus reviewed all 18 instances of adult neurosis that he had examined, and this labor was rewarded by a most significant discovery: the consistency with which the disturbed individual, often a woman, reported an instance of childhood seduction by a parent.

Further analysis revealed that this seduction had usually occurred by age three to five years. The same event experienced at a later age, even before puberty, did not result in a neurotic condition.

Freud was jubilant and with good reason. He had finally found the evidence, so long evasive, for Charcot's earlier contention that behind every neurosis lay some traumatic event. He had identified that event and even the time of its occurrence. He wrote to his only friend in these long years of isolation, describing the *seduction hypothesis* as his "momentous revelation," comparing it to the recent discovery of the sources of the Nile, about which there had been ignorance and dispute for thousands of years.

But the battle was not yet won. His profession ridiculed the idea and scorned his research. The seduction hypothesis encountered no support whatsoever, and one opponent openly referred to it as preposterous.

Freud emphasized that the statistics certainly were in his favor, for his findings included every case that he had analyzed up to that point. It was highly unlikely that all 18 persons constituted a series of odd exceptions to the way in which neurotic conditions developed. Had Little Hans appeared before him in these years, there is no doubt that Freud would have considered his case in this same way.

This research was based on the concept of *infantile sexuality*, which is the capacity of children to have sexual interests, an idea both unexpected and disreputable in his day. Sexual interests in children were considered impossible or abnormal, but the concept came to play a most prominent role in Freud's theory of personality.

During these difficult years Freud experienced his own adjustment problems, exhibiting some of the symptoms he had observed in others. Partly for these reasons and partly through scientific interest, he began an intensive effort at self-analysis. The doctor became his own most important patient, spending the last thirty minutes of each day in this effort. On going to bed each night he resolved to remember his dreams, and fifteen minutes after awakening he wrote down what he could recall. Later he engaged in free association about the contents of his dreams, and this effort was the most important part of his self-analysis.

One puzzling feature of all this work was that it did not reveal any seduction in his own infancy, despite his symptoms. It was entirely possible that he simply had forgotten the event, but this disconfirming case in his own life slowly began to undermine his whole conviction about the seduction hypothesis. If childhood seduction was the key, why was he unable to achieve complete success in therapy? If this memory of seduction was so important, why was it never recalled under other circumstances, outside of psychoanalysis? And, most important, how could it be that so many parents were prone to sexual perversions?

Gradually, the "awful truth" dawned on him; the seductions perhaps never occurred! Instead, his patients might have been reporting sexual fantasies which they believed or wished had happened earlier. Maybe there was nothing more, in most cases. His colleagues were right; the seduction idea had to be reconsidered.

With his hypothesis dashed, along with some hope of lasting fame, Freud went through a series of emotional states, including insecurity, resignation, and bewilderment. At one point he was ready to give up the entire goal, but later there emerged in him a growing conviction that his analyses, though somehow leading to the wrong answer, were nevertheless on the right path. The vaguely recalled sexual practices perhaps had not occurred, but the idea still might be significant in the person's thinking. Perhaps it was appropriate to stress childhood sexual desires and conflicts, rather than seduction itself. The fantasy constituted a psychic reality, whether or not seduction had actually happened. The seed of neurosis perhaps lay in this experience, fictitious or real.

Through these successive steps, forward and backward, at times *Theory of* hesitant and often ridiculed, Freud finally came to the position that *Psychoanalysis* infantile sexuality is the chief factor in the origin of adult neurosis. Around it he developed several related concepts and hypotheses that collectively constitute the theory of psychoanalysis. Much of this theory has remained constant since his day, though some aspects have been refined or rejected by those who followed him, attempting to carry on and verify his work. At the height of his career, as he awaited the arrival of Little Hans at 19 Berggasse,

Freud's *theory of psychoanalysis* was based on this issue of sexuality and three fundamental concepts: conflict, repression, and symbolic behavior. They form the foundation of psychoanalytic theory.

Conflict, the first of these, arises from the fact that life is difficult for everyone in one way or another. We all experience frustration, in which we are prevented from reaching a goal, and *conflict,* in which we are doubly frustrated because here two or more goals are incompatible. We experience opposing impulses or mutually exclusive goals. Little Hans, for example, wanted to strike his sister, but he did not want to be punished for this behavior.

Childhood conflict can be especially significant, for the habits, values, and attitudes adopted in these formative years can become models for subsequent development. One generally uses in later life those behaviors that have served earlier. For Freud, the first years are most decisive; the seeds of adult personality are to be found in the child's response to conflict.

Conflict, according to Freud, is most likely in sexual development, and in this respect psychoanalysis is essentially a biological theory. It argues that in childhood there is an inevitable unfolding not only of locomotion, grasping, language, and cognition but also sexuality. All these developments are significantly based in biological maturation. In learning to move itself from place to place, for example, the child progresses from lying to sitting to standing to walking, and a similar series of stages is evident in sexual development. There are of course individual variations in rate and mode of development, but the sequence is invariant in all normal persons.

These successive changes, in Freud's broad definition of sex, are called *psychosexual stages,* emphasizing that psychological development is closely tied to unfolding sexual interests. They begin with the oral stage, followed by the anal and phallic stages, and in each instance a characteristic task presents a special potential for conflict.

For the newborn baby, this task is the intake of food, which involves pleasurable stimulation of the lips and tongue. If the oral needs are adequately met during this first year, called the *oral stage,* then potential conflicts are avoided. The groundwork is laid for a basically optimistic and a relaxed outlook in the child. But if food is unavailable, if it is non-nutritious, if it is too much or too little, if it cannot be retained, if it is consumed too rapidly, if it leaves a gaseous residue, and so forth, these become grounds for a dissatisfied individual. Anxiety and pessimism have been fostered. Either way, the result becomes a model, or prototype, on which later responses are based.

In the second and third years, the basic task is focused around another erogenous zone, and this one caused Little Hans a bit of trouble. He fussed and stamped his feet in rage when interrupted from play for this purpose: toilet training. On occasion he threw himself to the ground and refused to leave what he was doing, and he admitted that he made a row when forced to go to the bathroom.

Like its predecessor, the *anal stage* involves pleasure associated with the alimentary canal. Tension develops in the body as waste materials accumulate, and the release of this pressure is satisfying. But this task is more difficult than that of the oral stage, for the child is requested to oppose its most basic nature, the inborn reflexes for elimination. And it is asked to do so by the very people who have loved and cared for it throughout its earliest years, a circumstance that adds to the potential for conflict. Toilet training represents the child's first extended confrontation with authority, and the management of this early learning, with its implications for conformity and disobedience, can have a lasting impression on the developing personality.

Around the beginning of the fourth year the average child reaches the *phallic stage,* during which it discovers the pleasures associated with stimulation of the genital areas. Freud referred to both sexes in this way, and in both instances these zones constitute a source of conflict for the child, as the emerging personality must learn what is punishable in this regard. Little Hans, just beginning this stage, asked his mother to stimulate him in this area. Much to his surprise she refused, explaining: "Because that'd be piggish."

"What's that? Piggish? Why?" the boy asked, trying to understand the adult restriction. Why should such a pleasure be denied?

The boy also had been told that Dr. A. came around on occasion and had a knife. It had even been suggested that he might use it in punishing such misbehavior.

Coupled with this growing interest in the genital area is another task—the further development of one's sexual identity, including a broader recognition of sex differences and a shift of interest beyond one's own body. In particular, there is in the phallic stage an intensification of affection for the parent of the opposite sex. This growing attachment, and an ensuing antagonism toward the other parent, also can have later repercussions, influencing the child's relations with others and its choice of role models. The ways in which these parent-child relationships are managed, according to psychoanalysis, are fundamentally responsible for the long-term outcomes in personality development.

Childhood conflict, then, is the first major concept in traditional psychoanalysis. It sets the stage for something to follow. Freud was careful to emphasize that conflict concerning sexual matters is only part of the formation of personality, but owing to its appearance so early in life, it is vitally important. The question it raises is this: What happens when the conflict cannot be resolved? How is the resulting anxiety managed?

Freud's answer here involved the novel idea of repression, the second major concept in psychoanalysis. In *repression* a person unconsciously excludes certain anxiety-provoking thoughts from awareness; the painful experience is kept out of consciousness. A child with sexual interests and fears of reprisal might repress the whole idea, thereby reducing or eliminating the anxiety.

Repression is only a partial solution, however. The conflict is not resolved and the anxiety does not disappear completely. Pushed to the realm of unconscious mental life, it is kept from daily awareness only at the cost of considerable psychic energy. Coping with the problem in the "back of his mind," a person's effectiveness in daily life is thereby reduced.

In dealing with repressed conflicts, Freud viewed his work like that of archaeology. Both the psychoanalyst and archaeologist explore the human past, excavating the human mind or a buried city. Piece by piece an earlier scene is restored, using only fragmentary evidence. Freud's consulting room, in fact, was overflowing with statues, busts, paintings, and bowls from all sorts of ancient civilizations, standing in half-ordered rows on the desk, table, shelves, and even the floor of his study. These had served the archaeologist in the reconstruction process, much as a patient's thoughts, memories, and dreams served Freud as fragmentary evidence. Carefully assembled and analyzed, they suggested some meaning in the patient's disturbed adjustment.

Freud's Relics. *This ancient Chinese scholar, flanked by Egyptian deities, was Freud's constant reminder of our buried past and the breadth of human experience.*

Repressed conflicts differ from buried relics in one important respect, however. They are more dynamic, for repression is not a fixed state of affairs. Any repressed thought may be expressed in a variety of indirect ways, and these disguises are called *symbolic behavior,* the third fundamental concept in psychoanalysis. Symbolic behavior, like any other symbol, stands for something else; it reflects conscious *and* unconscious concerns and thus may appear inexplicable to the casual observer. We unconsciously "act out" an underlying problem in indirect ways, such as through our dreams, in small but revealing errors, and in disturbed adjustment.

The outcome of this whole sequence—conflict, repression, and symbolic behavior—is known as *unconscious motivation*, which is the most significant and original contribution of psychoanalysis. It is the cornerstone of the theory, the piece that holds the rest together. Unconscious motivation means that human behavior can be influenced by earlier conflicts of which we are unaware.

Little Hans' behavior may reflect unconscious motivation. He is afraid of horses but cannot say why, and he has never been harmed by one in any way, as far as anyone can tell. He says he is fearful because horses may bite, but later he said he is afraid of horses wearing blinders and a muzzle, which of course prevents biting. Still later he says he is fearful because a horse may fall down, and when queried about these apparent discrepancies he simply says "fall down and bite." This perplexing behavior, according to psychoanalysis, may be a partial disguise for some underlying conflict, perhaps unknown even to the boy himself.

Oedipus Hypothesis At Hans' age, Freud hypothesized, the conflict is related to the phallic stage, particularly to the growing attachments and antagonisms concerning the parents. Freud called this emerging pattern of sexual interests the "romance of the family," but he also gave it the more formal title of *Oedipus complex*, a name taken from the early Greek drama. In this play young Oedipus, to his later horror, finds that he has unwittingly killed his father and married his own mother. The power of this tragedy, Freud claimed, is that the same potential curse has been laid upon everyone. It is the fate of every male child to direct his first inescapable sexual impulse toward his mother and his first hostile wish toward his father. The myth merely depicts fulfillment of this fate.

Every female, furthermore, is prone to the *Electra complex*, named for a young woman in Greek drama who was in love with her father and participated in the murder of her mother. Again the theme is assumed to be universal, and the conflict is considered to have long-standing personal significance. In these early exchanges of love and control, the social dimensions of the child's personality are formed.

The Oedipus hypothesis, together with the underlying idea of unconscious motivation, had been enunciated by Freud before he encountered Little Hans. They formed the basis of his approach to this and other cases. But a few years later he formulated still another trio of concepts that today are among his best-known ideas about personality. They help clarify the nature of individual conflict as he understood it, including the case of Little Hans. In examining this tripartite view of the human mind we must remember that Freud gave free rein to his imagination throughout his clinical work, presenting his findings in an unrestrained literary style.

There is in all of us, he stressed again, a certain biological inheritance, devoted to avoiding pain and obtaining pleasurable stimulation, such as food, warmth, sex, sleep, and so forth. Collectively, these desires are called the *id,* which is that part of the personality, totally inborn, committed to immediate gratification. The baby is said to be "all id," meaning that there is only this biological side to its personality. Little Hans wanted to have warm baths, to be caressed by his mother, and to be fed when hungry, all of which are part of the id.

The origins of the second dimension have been a source of speculation in psychoanalysis since Freud first promoted the idea, but it is generally agreed that the ego is partly an outgrowth of the id. The *ego,* which means "I" or "self," involves those functions by which the individual takes some voluntary action—perception, memory, thinking, and body movements. It is the problem-solving or executive dimension of the personality and functions in service of the id; its role is to find the best overall means of satisfaction, considering the demands of the id and the requirements of external reality.

The third and last component of mental life is something beyond the ego or self. It is therefore called the *superego,* referring to standards of behavior adopted from others. The time of its onset has been a source of debate, but it clearly emerges later than the ego, since it consists of those rules of conduct urged on the child by parents, teachers, and older siblings. Eventually, these ideas become important to the child regardless of the presence or absence of those adults who were most influential in their formation in the first place. The superego is thus capable of symbolically rewarding and punishing certain behaviors, many of which are not in harmony with the id.

Confronted with the requirements of the superego, the urges of the id, and the limitations in most human environments, the ego is faced with diverse and often conflicting demands. Sometimes it can handle them directly, as when Little Hans ventured into the country at his parents' request, despite his fear, but sometimes the problem is not so readily resolved and an indirect solution is employed. This indirect solution involves repression, followed by a variety of symbolic behaviors, as noted already.

This array of forces—the id, superego, and reality—constituted the components for conflict in Little Hans' case, presumably to be managed by his emerging ego functions. But did this conflict arise from the Oedipus situation? That was the basic question.

In Viennese society of Freud's day, the Oedipus hypothesis had special relevance, for a controlling father and tender, devoted mother were the expected parental patterns. Freud, in fact, developed this hypothesis chiefly through analysis of his own dreams, uncovering in himself an idea to which he would likely be most resistant: a deeply hidden, intense antagonism toward his own father. He saw this rebellion in a dream about an autocratic prince, in another about a young man who begged forgiveness from his father, and in a whole series about Rome, the "mother city." He also saw this reaction in mistakes in his writing, as when he wrote about Hannibal crossing the Alps and incorrectly referred to Hannibal's father as Hasdrubal, a half-brother who was later beheaded. In his own life Freud had a half-brother old enough to be his father; perhaps this slip of the pen represented both wishes, that a son could marry his mother and eliminate his father.

One can see in Freud's own life, as well as those of his Viennese patients, fertile grounds for ideas about the Oedipus complex, and eventually he decided that the difference between normal and neurotic people lay in the intensity of their unresolved feelings of love and hate for their parents. So when the case of Little Hans came to

Father and Son. *Jacob Freud was considerably older than his attractive wife and took responsibility for the boy's early discipline.*

his attention, he was immediately interested. Up to this time there had been no analysis of a child, and here was an opportunity to observe the early sexual problems directly, rather than through adult memory. Moreover, it would be an excellent occasion to test the hypothesis, for this five-year-old boy should be deeply immersed in the Oedipus struggle.

CHAPTER ELEVEN

Testing Hypotheses

"Will you come with me on Monday to see the Professor, who can take away your nonsense for you?" the father asked.

"No."

"But he's got a very pretty little girl," the man added, whereupon Little Hans readily consented. We do not know how he got along with the little girl, but today we remember her as the famous London psychoanalyst Anna Freud.

On the afternoon of the appointed day, the thirtieth of March 1908, the father and son arrived at the steep city roadway known as Berggasse, and at number 19 they passed through the foyer to Freud's waiting room and office, arriving promptly for his consulting hour. The doctor received patients regularly from three to four o'clock each day without appointment, as plainly marked on the doorway.

Freud had met "the little fellow" previously and knew him through the weekly reports from the parents. This information eventually came to occupy an important place in Freud's archives, as well as a prominent role in his clinical treatment of the boy.

This work with Little Hans also had broad significance for the clinical method. Coming as it did so early in the history of modern psychology, it gave direction to this approach. One of Freud's major contributions to psychology, apart from psychoanalysis, is the interest he generated in the clinical method.

Clinical Approach   In broad terms the *clinical method* includes the procedures for diagnosis and treatment of problems of personal adjustment. Sometimes it is called "the single case experiment," for the clinician usually works with one person at a time, and ideally each case is handled as a separate research question. The psychologist's first allegiance is to the troubled client, however, rather than to the acquisition of new information. For this reason there is typically less control than in experimental work and greater difficulty in generalizing the findings to other people.

The clinical method traditionally involves at least three major components. The father's extensive letters to Freud constituted a rough *case history*, which is the first component in most clinical work. It is a report of all significant details, past and present, bearing on the individual's adjustment, including daily routines, emotional reactions, successes and failures, and especially social behavior. The case history is usually compiled by a social worker, who talks with the family, relatives, and the individual concerned, but sometimes it is prepared by others, as in Little Hans' case, in which the father-physician played a prominent role.

The father, in a crude way, also played a role in the second element, *psychological testing*, which involves a series of questions designed to reveal the mental functioning of an individual. After Little Hans' giraffe dream, he suddenly asked the boy, "Just tell me quickly what you're thinking of." Hans' response was "raspberry syrup," which may have occurred for several reasons, as we will see later. In any case the father was attempting to use something like the *word association test* as a means of understanding the boy.

In this procedure a long list of words is read, one word at a time, and the subject responds in each case with the first word that comes to mind. The results are analyzed to discover what the subject says, how long it takes him to do so, whether he repeats responses, seems confused, and so forth. Experienced clinicians sometimes use such information in diagnostic efforts.

Sigmund Freud disdained the use of formal psychological tests, partly because they were less developed during his time and certainly because he considered the *clinical interview* as a psychological test, a commonly accepted idea today. In this interview, the third component of the clinical method, there is a conversation between the therapist and patient, and the emphasis is on understanding what the patient says and does in that relationship. Freud seemed to have a special capacity in this regard, and his interest in details and the concept of the unconscious led him to speculate about behaviors that often went unnoticed by others.

The traditional psychoanalytic interview would have been inappropriate with Little Hans, for at five years of age the boy would be reluctant to lie still on a couch, making efforts at self-discovery by concentrating on his thoughts and feelings. Instead, he and his father sat together with Freud and discussed the matter in normal conversation.

Freud's Desk. *Sitting in this chair, Freud talked with friends, colleagues, and prospective patients, including Little Hans.*

149

The interview was short and the only formal one the boy ever had with the founder of psychoanalysis. Apparently, Freud made this decision because he knew the case so well, had the father as an assistant, and appreciated the limitations of the boy's age.

The father began the conversation, explaining that Little Hans' problem continued despite his enlightenment about sex. The boy still wanted to stay with his mother and talked fearfully of horses. He had even become afraid of other large animals and would not readily venture outside. Only his relations with Hannah seemed improved.

Freud looked and listened. The father and son sat there before him discussing the problem, but the pieces of the puzzle still did not fit together. The father emphasized that Hans was particularly bothered by horses with black around their mouths and wearing blinders next to their eyes. These factors were unexplained. Suddenly, as he listened to the parent and heard the boy's own description of the problem, Freud was struck by an insight which, he reported later, might readily have escaped the father.

Jokingly, Freud asked Little Hans whether the horses in question wore eyeglasses, and the boy replied in the negative. He then asked if his father wore eyeglasses, which Hans again denied, despite his father sitting right next to him with eyeglasses on his face. Thereupon Freud asked whether the blackness around the horse's mouth perhaps indicated a mustache. And, eventually, he arrived at these thoughts with his visitors: perhaps Hans' fear of horses was partly a disguised fear of his father, which arose partly because of the boy's fondness for his mother.

The father was big and powerful, as indeed are horses. He had dark objects around his eyes, similar to the horse's blinders, and a darkness around the mouth, similar to the muzzle. Then, too, it was the father's custom to give Hans horsey rides, at which times the boy could feel the man's strength. In the street the boy had seen horses pulling heavy loads, and he said he was afraid they would fall down, a fear he also experienced when he rode about on his father's back.

It must be, Freud explained to Little Hans, that he thought his father would be angry with him because of his yearning for his mother, but his father was fond of him anyway and Hans might tell everything without fear. Long ago, Freud continued, he had known that Little Hans would come into the world loving his mother so much that he certainly would be fearful of his father.

The father interrupted, asking the boy, "But why do you think I'm angry with you? Have I ever scolded or hit you?"

"Oh, yes. You have hit me," Hans responded.

"That's not true. When was it, anyhow?"

"This morning," explained the boy, and then the father recalled that Hans quite unexpectedly had butted him in the stomach, with the result that he struck his son in reflex fashion.

The father apparently did not fully recognize his son's competitive disposition toward him, which prompted the ensuing fear that Little Hans wished to disguise. The horse was a suitable substitute, for it symbolized the man in several ways. It was also relatively harmless and could be more easily avoided. Hans had little to gain by showing his resentment directly, and besides he loved the man, who had been helpful in many ways.

This childhood conflict of love and hate had been repressed and reappeared in symbolic form. The disguise allowed the boy to love his father more completely, and by this same means he could seek still further contact with his mother. His fear of horses was a symbolic effort to manage these diverse implications of the Oedipus complex.

All of this was not explained to Little Hans, but he was duly impressed anyway. On the way home he asked, "Does the Professor talk to God. . . ?"

But Freud was less satisfied, feeling that this observation was not the sole key to the problem. He hoped it would prompt other ideas and called his interpretation only a "piece of the solution." It did, however, offer the boy the possibility of bringing forth his underlying feelings, thereby further unraveling the phobia.

Among the various possibilities for helping someone by means of *Dream* psychoanalysis, Freud also placed great stock in the study of dreams, *Interpretation* which he called the royal road to the unconscious. In the *interpretation of dreams* the analyst attempts to understand some problem of adjustment by determining the underlying meanings of that person's dreams. Again the three-step model of unconscious motivation is involved, but it is approached in reverse. One begins with the dream symbols, attempting through psychoanalysis to lift the repression, thereby revealing the conflict.

Little Hans had the following dream: In the night a big giraffe and a crumpled one were in a room. The big one called out when Hans took away the crumpled one. Then it stopped calling. And then Hans sat on the crumpled one.

This report is the *manifest content* of the dream, which is simply the dream story; it is what Little Hans was seeing and hearing in his sleep. Freud suggested that the manifest content is heavily influenced by activities of the preceding day or two, and modern research supports this view. Dreams usually involve recent experiences, or memories evoked by recent experiences, and just before this dream Little Hans had seen giraffes at the zoo. But according to Freud the manifest content by itself is not particularly important, except as it is symbolic of the underlying problem.

Beneath these surface fragments lies the true significance of the dream, called the *latent content* because it is hidden. This content concerns unconscious problems in the dreamer's life, and the aim of dream interpretation is to discover these hidden concerns.

Freud at one point decided that the latent content of virtually all dreams has the same theme, and in taking this stand he knew he would meet with the most emphatic contradiction. But he said it anyway: "The dream in its inmost essence is the fulfillment of a wish." In fact, he was so confident in this idea, which came to him most fully one day on the patio of Bellevue Hotel, that he jokingly suggested a tablet should commemorate the spot where "the secret of dreams was revealed to Dr. Sigmund Freud."

Usually the wish is concealed and, if it is important, it has roots in childhood. What then was Little Hans' concealed wish, which certainly would come from childhood?

The answer to this question requires an understanding of the *dream work*, by which the underlying wish, or latent content, is transformed into the manifest content. In Freud's view the basic process in dream work is the use of symbols. If the latent content involves a male, for example, a pole, stick, dagger, or some other elongated object may appear in the manifest content. The male is not necessarily present; this person is merely represented by a symbol of male sexuality or by some other disguise.

Similarly, the woman might be represented in the form of a cave, jar, building, box, or similar receptacle, considered symbolic of the female. In a like manner images involving water are thought to suggest an underlying theme of birth or life, for human life first appears in this context and water is necessary for life. Freud spent considerable time justifying his view of such symbols as universal, pointing to their origins in language and custom in widely different cultures.

These symbols, Freud explained, contribute to the riddle of dreams, sometimes making them confusing, but they also increase the dream's efficiency. There never were any real giraffes in Little Hans' room; giraffes do not call out; they are not crumpled; and he never sat on one. But this puzzling dream is a condensed expression of Little Hans' childhood wish, symbolized through animals, which often portray human beings in children's stories.

In this wish, according to Freud, the big giraffe, large and strong, represents Hans' father, its elongated neck symbolizing the male. The crumpled giraffe is his mother, smaller in size and with less obvious genitalia. The fact that the animal is crumpled but still somewhat lengthy may be accounted for partly by the disguise process of dream work and partly by the fact that Hans, despite many requests, had not seen his mother naked. He thought she possessed an organ like his own, yet he also imagined it to be something like Hannah's.

On the night of this dream, March 27, the boy surprised his parents by coming into their room when it was still quite dark. Entering their bed, he fell asleep, after which he was carried back to his own room.

According to the father, this was a common scene in the parents' bedroom; the boy often arrived there with the purpose of entering the bed with his mother. The father cautioned his wife about submitting to this request, but she usually replied that he need not be concerned. A moment or two with Hans in bed could be of no significance. During the summer holidays, when Hans' father was away from the family, the boy frequently slept in his mother's bed.

The dream is a representation of this event, according to Freud. Not wanting the boy to enter the bed, the symbolic father calls out, and when he stops calling, Hans sits on the crumpled giraffe. This response is his effort at taking possession of his mother, a reasonable simulation considering his lack of information about sex and the disguise process of the dream work. Little Hans in this dream is a little Oedipus, wanting to possess his attractive mother.

The giraffe dream is thus a continuation of the boy's fear of horses, a transposition of the Oedipus wish into giraffe life. The wish appeared during a dream, according to psychoanalysis, because the ego is less vigilant while a person is sleeping. A forbidden impulse can more readily pass the barrier of repression—at least in partial disguise. This dream was considered by Freud as further evidence for the Oedipus hypothesis.

The dream of the plumber who removed the bathtub can be regarded in the same way. Hans was alone with his mother, bathing in contented fashion and suddenly a man with a large borer entered the bathroom. He took away the means by which the little boy was enjoying an intimate relationship with his mother, and then he became aggressive toward the child, sticking a borer in his stomach.

Even the boy's dream of crawling under the ropes at the zoo, precipitated partly by events of the previous day, can be regarded in this way. Hans ignored the prohibition of the rope and then was apprehended by the police. Later that same day he confessed another impulse: "I went with you in the train, and we smashed a window, and the policeman took us with him."

In both fantasies the boy is expressing a forbidden wish, seemingly symbolic of the incest taboo, and in his imagination he overcomes the restraint. Smashing a window allows entry, just as crawling under the rope permits entry into an enclosed, forbidden space, representing the female. Apparently Hans perceived his father as sharing this impulse, for in the latter instance both of them were taken into custody.

According to psychoanalysis, each of these dreams includes disguised elements of the Oedipus complex that passed the barrier of

At the Zoo. *Visitors were kept at a safe distance from the animals at Schönbrunn, though a small boy certainly could crawl under the fence.*

repression. Together with observations from the interview, they left Freud so confident of his interpretation that he said he learned nothing new from the case; it simply confirmed what he already knew. But he was pleased, for the case gave "more direct and less round-about proof" for his hypothesis.

He stressed that symbolic behavior usually has more than one set of determinants. The giraffe dream presumably was influenced by the visit to the zoo and possibly the picture over Hans' bed, as well as the father-son rivalry. The dream of crawling under the barrier had its origin in the roped-off area, the father's remark, and also elements of the Oedipus struggle. Freud referred to this phenomenon as *overdetermined behavior*, meaning that a response can be influenced by several factors. He decided that most behavior comes about in this way and sought to understand the relationships among the relevant factors through exceedingly detailed analyses.

Overdetermined Behavior

There is, for instance, the still unanswered question concerning Little Hans' special fear of heavily loaded horses, which Freud considered in this same way. Hans had seen one fall and it gave him a fright, but according to Freud this event, if significant at all, was only a contributing cause of the fear. It had been sustained and augmented by other factors. Just before birth, for example, the mother is always heavily loaded, much like a loaded horse, and falling down perhaps was something like having a baby. Hans knew that blood was involved, as in falling down, and the outcome, according to his experience with Hannah, was highly undesirable.

Once the father and son reversed roles, and the man asked where they got Hannah. The boy replied, "In the box; in the stork-box." Later he claimed: "Even while she was still traveling in the box she could run about and she could say 'Anna.'"

These diverse incidents strengthened Freud's conviction that in the boy's mind horses pulling furniture vans perhaps had some relationship to that great and perhaps fearful riddle of humanity, the birth of a baby. Once when Hans saw horses pulling carts loaded with luggage, he inquired, "When Mummy was having Hannah, was she loaded full up too?"

And at this point we have come full circle in examining Little Hans' disorder. According to psychoanalysis, he developed his fear because horses represented his father, because the fear allowed him to stay with his mother, because of the warning against horse bites, and because he had seen a loaded horse fall down, which apparently had some connection with his mother's pregnancy. His illness, the rivalry with Hannah, and the conflict between his parents also may have contributed to this behavior. Among all these circumstances the Oedipus complex was the precipitating factor, but it occurred in the context of several predisposing conditions.

In this same way simpler responses also can have more than one determinant. After questioning Hans about the giraffe dream, the father asked, "Just tell me quickly what you're thinking of." "Raspberry syrup," the boy replied. This substance was used for his constipation, but perhaps the full response was also related to blood, the birth of Hannah, and even Hans' unacknowledged anger at his father, not only for the man's deceit about birth but also for the Oedipus situation and his insistent questioning.

The father persisted, "What else?" And he received a terse reply: "A gun to shoot people with."

In the days following the interview with Freud, the father con- *Play Therapy* tinued to walk and talk with the boy. He often inquired, "What did you think of?" "What did you do then?" "What do you remember?"

Once after they had been out walking Hans told an elaborate story about seeing his father lay an egg in the grass. Later a chicken came hopping out, and Hans' mother saw it, too. The father protested that the episode was untrue. He would confirm it with his wife.

"It isn't true a bit," Hans replied. "But *I* once laid an egg, and a chicken came hopping out."

"Where?" asked the father.

"At Gmunden I lay down in the grass—no I knelt down—and the children didn't look at me, and all at once in the morning I said: 'Look for it children; I laid an egg yesterday.' And all at once they looked, and all at once they saw an egg, and out of it there came a little Hans. Well, what are you laughing for? Mummy doesn't know about it, and Karoline didn't know, because no one was looking on, and all at once I laid an egg, and all at once it was there. Really and truly. Daddy, when does a chicken grow out of an egg? When it's left alone? Must it be eaten?"

The father explained about chickens and eggs, and Hans concluded: "All right, let's leave it with the hen; then a chicken'll grow. Let's pack it up in a box and let's take it to Gmunden."

Two days later the father pursued this subject again. He explained that children grow inside their mothers and are brought into the world by being pressed out of them. He added that pain is involved.

The next day he asked, "Can you remember how the cow got a calf?"

"Oh, yes. It came in a cart," Hans replied, using an explanation that probably came from Gmunden. "And another cow pressed it out of its behind," he said, apparently trying to reconcile the cart idea with what he had learned from his father in the previous conversation.

In any way possible the father prompted Hans to divulge what was on his mind. According to Freud's instructions, he was prepared to offer further clarification of the Oedipus conflict, awaiting only the boy's readiness. On one occasion the father and son even went near the Hauptzollamt station, where horses frequently appeared. The father also joined Little Hans at play, and one day they

Hauptzollamt Station. *This railroad and customs depot could be seen from Little Hans'*
house.

began to discuss horses. Hans said: "The horses are so proud that I'm afraid they'll fall down." He seemed to be referring to the prancing coach horses, which held their heads high, but the father asked who it was that was proud. The boy said: "You are when I come into bed with Mummy." The father inquired, "So you want me to fall down?" Hans replied: "Yes. . . . When you come up to our flat, I'll be able to run away quick so you don't see."

In these ways Freud and the father used a necessarily modified approach to therapy with Little Hans, who was much too young for traditional psychoanalysis. Rather than talking directly about their problems, young children prefer to express themselves in some form of play and this remedial procedure, called *play therapy*, can be employed in a clinic or at home. Little Hans played at home with dolls, loading and unloading toy carts, and at times he pretended his father was a horse.

In a typical session various toys are made available and the child is allowed to do whatever he wishes, providing there is no personal injury or destruction of property. He may dramatize a situation from real life or produce a story from imagination. It is assumed that these activities are in some way meaningful to the child, as when Hans once described a fantasy about whipping horses.

This free play with toys is comparable to the adult's free association with words. Feelings are expressed spontaneously, and the therapist provides opportunities for further communication, developing a close, accepting relationship with the individual. The father permitted Little Hans to play in these ways, "interviewing" him at judicious moments and adopting whatever role seemed most appropriate at the time.

One morning when the man was preparing to leave the breakfast table, his son said: "Daddy, don't trot away from me!" Surprised by the word "trot," the father replied, "Oho! So you're afraid of the horse trotting away from you." The father had decided that the boy was ambivalent about him, for along with Little Hans' fear and resentment there was also a great deal of affection.

Gradually, as the boy was encouraged to express himself, and perhaps because the father was spending more time with him, there was further improvement. Previously, Hans had run into the house at the sight of an approaching horse, but weeks later he stayed in the doorway for as long as an hour. He shrank back no more, despite the passing of horse-drawn carts. Still later, he occasionally ran into the street, and afterwards he took regular walks with his father.

On the Promenade. *Little Hans and his father might have looked like this man and boy, strolling near the Opera House.*

Indoors he even adopted the role of what he had feared most, playing that he was a horse. He trotted, fell down, kicked his feet, and neighed, and sometimes his father played with him. The boy once tied a small paper over his face, like a nose-bag, and on several

160

occasions ran up to his father and bit him. When his father persisted with questions, the boy requested impatiently, "Oh, do let me alone."

One April morning Hans came into the bathroom while his father was washing. The man was naked to the waist and his son observed: "Daddy, you *are* lovely! You're so white."

"Like a white horse," the man suggested, seizing this opportunity to prompt the boy.

"The only black thing's your mustache," Hans responded. Then, after a moment, he added: "Or perhaps it's a black muzzle?"

Toward May the progress was unmistakable. Hans never shrank back and when taken for walks he chased after horse-drawn carts. A concern about going into other neighborhoods was the only vestige of his former fear. The Oedipus struggle apparently was being resolved partly in fantasy and partly in a realistic acceptance of the situation.

When Hans was playing with some imaginary children, his father asked: "Are your children still alive? You know quite well a boy can't have any children."

"I know. I was their Mummy before, now I'm their Daddy."

"And who's the children's Mummy?"

"Why Mummy, and you're their Grand-daddy."

"So then you'd like to be as big as me, and be married to Mummy, and then you'd like . . . to have children."

"Yes, that's what I'd like, and then My Lainz grandma would be their granny."

Hans in these ways seemed to be solving the problem through identification with his father. In the process of *identification* a child adopts the behavior of the like-sexed parent, rather than struggling against that person, thereby gaining the characteristics of that individual to some degree. A boy typically identifies with an adult male, especially his father, and a girl follows the examples of her mother or another older female. Identification, often a long process, is considered one of the most effective means of resolving the Oedipus-Electra struggle, especially in normal child development.

Hans' play with his children perhaps was in service of both these wishes—to replace his father and also to be like him. For instead of

161

destroying the man in this fantasy about marrying Mummy, Hans bestowed on the man the same circumstance that he desired for himself. The father, as the imaginary children's grandfather, would be married to the Lainz grandma, his own mother.

CHAPTER TWELVE

Verification

So Little Hans was a phobic boy who proved to be not so phobic. He might lead a horse to water and he would not even shrink. At least he stood his ground in the presence of horses and seemed more normal to Freud, who was an astute observer of human behavior and well versed in the clinical method.

In the spring of 1922 Freud answered a knock on his door and received a great surprise. A youth said, "Ich bin der kleine Hans," meaning "I am Little Hans." The young man was on his doorstep for a brief visit. He appeared strong and healthy and seemed without personal difficulties, but the more remarkable feature of his visit was his memory. He could not recall his fear of horses or any of the events leading up to it. He read his own case history and, except for the Gmunden incident, none of it seemed familiar. He did not even recognize himself. At nineteen years of age he had forgotten his phobia and the entire effort at psychoanalysis by Freud and his father. This forgetting, according to Freud, was further evidence for his ideas about unconscious processes, and the boy's healthy condition offered support for psychoanalysis as a method of therapy.

But these claims need verification. Without it, they do not meet the requirements of science.

Freud never saw Little Hans again, and the boy was not studied by any other clinician, which leaves the case in doubt. By itself it is not of fundamental significance, however, except as a landmark in psychoanalysis. More important are the issues it raises concerning infantile sexuality and the Oedipus complex, for which further support is needed.

Modified Psychoanalysis The first place to search for support is elsewhere in psychoanalysis and, except for Freud, the most noted psychoanalyst has been Carl Gustav Jung. A much younger man, his name is pro-

Carl Gustav Jung. *Planning a career in archeology, he was prompted partly by his dreams to consider biology and then psychiatry.*

164

nounced that way, like "young." Also interested in childhood sexuality, he was regarded by most early observers as Freud's heir-apparent in psychoanalysis. Both men studied dreaming, and they came to America together to discuss their work. During the voyage they passed several hours analyzing each other's dreams. Their growing popularity prompted James Joyce, the Irish novelist, to quip: "Americans are jung and easily freudened."

In his lectures, Jung spoke on his recently completed analysis of a four-year-old girl, Little Anna. Certain details of this case, he said, gave clear evidence of infantile sexuality. In fact, the analysis showed "astonishing similarity" to Little Hans' story, though it had been handled independently, without any knowledge of Freud's work. In this way, it verified Freud's finding.

Little Anna was lacking information about sex and much concerned about the origin of her younger sibling, Freddy. Like Hans, she asked her parents about childbirth several times, and one day she tried to find out from her grandmother.

"Grandma, why have you such withered eyes?"

"Because I am old."

"But you will become young again."

"No," replied the woman, "I shall become older and older, and then I shall die."

"Well, and then?"

"Then I shall become an angel."

"And then will you become a little child?"

Anna apparently was attempting to reconcile the story of the stork, related by her father, and the question of death. Her understanding, which received no correction from the grandmother, was that babies are little angels from heaven, brought to earth by the stork. After they die they go back to heaven to live as children.

She was also struggling with the problem of her new rival. The evening before Freddy's birth her father had asked, "Tell me, what would you say if we should get a little brother tonight?"

"I would kill it," came the prompt reply.

A few days later at dinner Anna announced that she had had a dream. It was similar in content and brevity to that of Little Hans:

"I dreamed last night about Noah's ark, and there were lots of animals on it, and underneath there was a lid that opened and all the little animals fell out."

Anna possessed a toy Noah's ark, which may have been partly responsible for the manifest content of the dream, but its opening was in the roof, not the bottom. The underlying meaning, Jung decided, perhaps reflected her increasing awareness of childbirth. She had learned that the newborn somehow emerges from underneath the mother, not from the mouth or a hole in the breast, as she had been told.

Later, she reported another dream: "I dreamed about a woman in the city; she had a very big belly." Anna apparently was making progress.

While working on this question, she took pleasure in helping the gardener plant grass. The significance of this activity for the child was probably far greater than the man realized, and later she asked her father: "Tell me, how did the eyes grow into the head?"

"They did not grow into the head, they were there from the beginning. . . ."

"Were not the eyes planted?"

"No," he answered, "they grew in the head like a nose."

"Did the mouth and ears grow in the same way? And the hair too?"

"Yes, they all grew in the same way."

"And the hair too? But the mousies come into the world naked. Where was the hair before? Were there no seeds added?"

"No," the father explained. "You see the hair really came out of little grains which are like seeds, but these were already in the skin before and nobody sowed them."

Finally Anna asked in a stressful tone: "But how did Freddy get into Mamma? Who stuck him in? And who stuck you into your mamma? Where did he come out from?"

These very precise questions could hardly be avoided without further deceit, and they prompted a fuller explanation from the father. He described himself as a gardener and the mother as the soil of the earth. Their roles lay in planting and nurturing the seed, respectively, giving birth to the baby. Apparently he included sufficient detail to satisfy the child, who listened with fullest attention. Then she ran to her mother and said: "Papa has told me everything; now I know it all."

The next day she went again to her mother and said: "Think, mamma, papa told me how Freddy was a little angel and was brought from heaven by the stork."

The surprised mother replied, "No, you are mistaken, papa surely never told you such a thing!" Whereupon Little Anna ran off with a laugh, apparently in revenge, Jung suspected, for the falsehoods told to her earlier.

Jung concluded that the case provided verification for the concept of infantile sexuality. The idea that a child could not possibly be interested in sex was a myth contradicted by Freud's clinical findings and further dispelled by his own work. Little Hans was not alone, and shortly thereafter legions of similar reports followed, adding further verification. Today, amid all this evidence, there is really no significant debate on this issue. Teaching children facts and attitudes about sex, called *sex education*, is an outgrowth of these studies and a widely accepted practice.

There are, however, differences of opinion about how the flight of the stork should be presented. Which details should be included? How should they be taught? And when? Should this teaching be the exclusive right of parents, the school, or a shared responsibility?

Such questions are beyond the scope of our study, but one guiding principle perhaps can be derived from instruction in other school subjects. Mathematics, language, history, and science are all begun early, with no attempt to teach everything in the first grade. Instruction starts with the simpler elements and year by year greater complexity is added. Little Anna ran to her mother saying: "Papa has told me everything; now I know it all." But that was not true. The topic of sex is broad indeed, containing biological, psychological, and social dimensions that can hardly be considered adequately in one lecture, one course, or even one year. And since all of us are always in the process of change, especially in early childhood and adolescence, the teachable moment for the different units in sex education will vary from age to age.

Anna's insistent questions furnished evidence for childhood sexuality, but they offered no convincing support for the Electra hypothesis and no such details were mentioned by Jung, an omission that is noteworthy in view of Anna's age. The case suggests either that Jung was not thorough in his analysis or that the hypothesis does not have universal application.

Even after he accepted a great deal of Freud's thinking, Jung did not place much credence in the Oedipus-Electra hypothesis. For him it was merely an assumption and usually not justifiable. The incest problem, he said, "signified a personal complication only in the rarest cases." Jung believed that the Oedipus complex, more than anything else, was a personal concern for Freud.

After some years he and Freud came to a sharp disagreement over the role of sexual factors in personality development, and each went his separate way. Freud remained with his approach, known as *traditional psychoanalysis,* and Jung developed another view of psychoanalysis, called *analytical psychology.* Infantile sexuality is recognized in analytical psychology, but it is not the chief factor in adult personality disorder. Current conflicts are emphasized, rather than earlier ones, and the Oedipus hypothesis is not central to the theory.

A man with enormously diverse interests, Jung's analytical psychology indeed has a broad scope. It includes concepts from many different fields and all sorts of civilizations, ancient and modern, Eastern and Western. Jung also stressed that each individual's mental makeup is composed not only of his or her own personal unconscious but also the remnants of experiences of all our ancestors throughout human and animal history. This *collective unconscious* reflects our racial and animal past and is said to be inherited by all members of the species. Not surprisingly, it is regarded with some skepticism in certain quarters.

Similarly, the work of Alfred Adler is widely recognized in psychoanalysis, and it too should be considered with respect to the findings in the case of Little Hans. After hearing a lecture by Freud in 1902, Adler became a supporter, then a disciple, and finally co-editor of Freud's journal of psychoanalysis. His writings therefore should reflect comparable interests, and perhaps they provide support for the Oedipus complex.

The Adlers. *With his psychoanalytic views and her emancipated style, they were an exceptional Viennese couple.*

In approaching Little Hans, however, Adler would turn instead to the previously mentioned concept of sibling rivalry, for which he is responsible. He would note the boy's competition with Hannah, emphasizing that the way in which this rivalry was worked out, beginning with Hans' concerns about the birth of the baby, would be most influential in determining his personality as an adult. The chief source of aggression in Little Hans' life, according to Adler, was not so much competition with his father over his mother as competition with his sister over both his parents. For Adler, too, evidence for the Oedipus-Electra hypothesis was not impressive.

Eventually, Adler also separated from Freud and developed his own modified approach to psychoanalysis. Called *individual psychol-*

ogy, it emphasizes that each person is to be understood in terms of the individual goals for which he or she is striving. The origins of these efforts, he speculated, lie in some physical deficiency, which virtually everyone has in one form or another. In our attempts to overcome this feeling of deficiency, we compete with peers, and in this context Adler coined another term of widespread interest. An *inferiority complex* is the sense of inadequacy which presumably most of us experience in relation to other human beings, prompting our efforts at individual assertiveness.

In the long-term development of traditional psychoanalysis one can see that aggression plays a gradually increasing role as a companion to the sexual impulse, and Adler is partly responsible. He influenced Freud's thinking in this way, just as his own thinking had been influenced by Nietzsche's doctrine of will for power. But compared to Freud's psychoanalysis, there is relatively little concern for sexual factors and unconscious processes in individual psychology.

Still other psychoanalysts and clinicians suggest that Little Hans' phobia arose over tensions in the family and concerns about life and death. They offer a variety of views about Little Hans and the Oedipus-Electra hypothesis in general, for which support is equivocal. In recent years, in fact, the Oedipus issue has been somewhat displaced as the chief concern in psychoanalysis. There is instead increasing interest in the pre-Oedipal years, from birth to age two or three, and in the identification process during the latter part of this period.

Anthropological Research   If the Oedipus-Electra hypothesis is universal, it should appear in cultures around the globe, regardless of the theoretical views of Freud's associates. Hence, soon after Freud announced his findings in America, the search began. Evidence was sought on all continents, and the basic question was this: Would the Oedipus complex be found in a family structure very different from the one predominating in Western Europe in the late nineteenth century?

This question involves *cross-cultural research*, in which societies are compared with respect to cultural practices. The aim is to study the effects of one or another custom by noting characteristics associated with its presence or absence. The difficulty, of course, is that societies often differ in many ways, apart from the practice in question; hence, the influence of a given cultural practice sometimes cannot be determined precisely.

Several celebrated cross-cultural investigations were accomplished by European and American anthropologists in the Trobriand Islands in the 1920s. For years they compared these family patterns with those of Western Europe for the same era, and eventually two major points of difference were identified. First, much greater sexual freedom was afforded children in the Melanesian society. Young boys and girls engaged in the most diverse forms of sexual expression without being regarded as reprehensible. Many of their games, played in the middle of the village apart from adult supervision, included a sexual or sensual element. Such pastimes were regarded by their elders as innocent amusement, as the partici-

Melanesians at Play. *Their games are observed here by Bronislaw Malinowski, a European anthropologist.*

pants initiated one another into the mysteries and practices of sexual life. Even in the home there were no special precautions to prevent offspring from witnessing their parents' sexual behavior. The child was simply scolded and asked to pull a mat over his head.

Second, the power within the family was much greater for the father in Europe and for the mother in Melanesia. The reckoning of kinship among the Trobrianders was matrilineal, with succession and inheritance determined according to the female line, and the children were members of the mother's clan and community. The father's role in childbirth was not even recognized. Children were believed to have been placed in the mother's womb as tiny spirits, and after the birth the father became the child's beloved companion and caretaker, but not disciplinarian. That role, chief of the household, was left to the wife's brother. The typical Trobriand father did not have the almost exclusive "right to the mother" found in the European family.

One apparent outcome of these differences in family practices concerned the latency stage. In psychoanalysis the *latency stage* is that period of late childhood during which there are no manifest sexual interests. In Western society it generally begins around age six, immediately after the phallic stage, and lasts until puberty, around age eleven or twelve, at which time sexual interests rise sharply. In other words, the sexual impulse coincides with physical growth, according to psychoanalysis. It is evident during the rapid changes of early childhood, temporarily decreases in the latency stage, and revives with the growth spurt of adolescence, when the young adult begins to form mature interpersonal relationships of a sexual and more general nature.

But sexual interests did not disappear in late childhood among the Trobriand Islanders. They continued without any latency stage, apparently due to the more permissive atmosphere, and there was less friction between the father and son, apparently due to the different family power structure. The aim of the Oedipus complex, if it existed at all, was "to marry the sister and kill the maternal uncle."

Young adults preparing for marriage were subject to certain taboos, nevertheless, some of which are not typically experienced in

Western society. Several couples lived in the center of the community in a "bachelor's hut," where they shared a place in this building but never a meal. For a young man to invite his girlfriend to dinner in the tradition of Western society would have led to her utter disgrace.

Bachelor's Hut. *The choice of partner in any such an arrangement, according to Freud, is influenced by the earlier Oedipus-Electra relationship; some fragment of the opposite-sexed parent is perceived in the new loved one.*

The Oedipus complex did not appear in the same form and intensity as in Western Europe, and investigations among the Hopi Indians of North America yielded similar findings. Personal relationships seemed dependent upon particular features of family life and sexual mores of that society. Among the Hopi in the first half of this century, the Oedipus complex seemed essentially absent.

Those in favor of the hypothesis had a rebuttal. The data were perhaps correct but not the interpretation. The basis for the Oedipus struggle was still present among the Trobrianders and Hopi, but it had been diverted into a similar problem among different family members. Furthermore, the traditional Oedipus complex was evident in other civilizations: among the Fan of West Africa, in the epics of ancient India, and in Middle Eastern civilizations. If one looked in the right places, or perhaps with the right frame of mind, it was readily available.

What was needed was an integration of all the cross-cultural studies, and for this purpose comparisons were made among many societies in Asia, Africa, and the Americas. These data were collected by many different researchers, but the results were still inconclusive. They showed the Oedipus complex to be "approximately valid" for young boys attracted to their mothers, but otherwise the findings were not readily interpretable.

Experimental Studies Mixed findings are not unusual in cross-cultural research; many relevant factors remain uncontrolled. For this reason some investigators, in the manner of Oskar Pfungst, transfer their studies to the laboratory, which permits greater precision. In this mode of research, the experimental method, the investigator selects the research subjects and tasks for a specific purpose, such as a controlled test of the Oedipus-Electra hypotheses.

In one instance many young women were studied for their attraction to male physiques, which according to the Electra complex should resemble those of their fathers. Series of photographs were assembled showing all sorts of males: thin men, muscular men, and heavy men. Each woman was asked to choose from several assortments the physique she would most prefer in a lover, and the results were happily received in psychoanalysis. The subjects tended to select body types that were most like or least like those of their fathers. These findings, according to the theory, suggest a strong sexual attachment to the father figure or an unconscious struggle against this attachment, both supporting the Electra hypothesis.

But there are obvious limitations in this research design. The term *research design* refers to the overall plan or scheme of an inves-

174

tigation; it includes the selection of subjects, development of apparatus, and procedures by which the data are to be collected. The research design in this instance included only women; it measured just preferences in photographs, not real life; and, most important, there was no concern with personal relations of any sort. The research is precise, and therein lies its merit, but the reasons why the women made their selections are merely speculated, not demonstrated.

This research also illustrates a general limitation in Freudian theory. Whenever a woman selected a physique resembling that of her father, the result was regarded as indicating a sexual attachment to that parent. Whenever a woman chose a physique quite unlike that of her father, this choice was viewed as reflecting an unresolved Electra complex. Psychoanalysis is supported either way, which leads to the criticism of imprecision in the theory, though in fact it may even be correct.

Hundreds of children between five and sixteen years of age served in another experiment, performing a variety of activities designed to test the Oedipus hypothesis. They completed unfinished fables, studied ambiguous pictures, and wrote stories about the pictures—all with a potential Oedipus theme. When these results were analyzed, it was found that the boys' stories indicated more positive impulses toward the mother and more negative impulses toward the father than did those of the girls, thereby supporting the Oedipus hypothesis.

But there are also limitations in this research design. The results concern themes in childrens' stories, not actual responses toward a parent, and they do not necessarily reveal a latent Oedipus complex. Boys and girls differ in many ways at various ages, and mothers and fathers differ in their management of children. Parental efforts at authority and control, apart from sexual attraction, could influence these outcomes.

The experimental method can provide a precise test of a hypothesis, but the data are collected under contrived conditions, sometimes quite different from the blooming, buzzing confusion of daily life. The findings may be precise, but they are not necessarily decisive.

These diverse results from clinical, anthropological, and also experimental research point to one outcome: evidence for the Oedipus-Electra complex remains inconclusive. The idea is an *unverified hypothesis*, which does not mean that it is incorrect or disproven but rather that more convincing data are needed. Psychoanalysts are in some disagreement; anthropologists emphasize that the form and extent of the complex depends upon the family structure; and experimental studies give only qualified support, which also suggests that the question is not well suited to contemporary research methods.

So the hypothesis remains a hypothesis and one which is almost untestable, based upon such highly abstract concepts as unconscious motivation and symbolic behavior. These concepts can be approached and defined in several ways, leaving many possibilities for disagreement, mixed findings, and vague claims.

In Freud's work with Little Hans the chief support for the Oedipus complex comes from Freud himself. There is no direct, independent evidence that Hans actually did fear his father and desire his mother sexually. Freud assumed that this was the case and proceeded accordingly, but even in Freud's data Little Hans never showed any outright fear of his father and never said he was afraid of him.

Signs that the child wanted increased contact with his mother are somewhat more substantial. The boy expressed a desire to be with her, to "coax" with her, and to come into her bed, though this behavior does not necessarily demonstrate that he also wanted sexual relations.

There is also Hans' alleged improvement during Freud's work on the case, an interesting but inconclusive piece of evidence. Simply growing older, moving to another house, having a new maid, and talking more frequently with his father may have influenced this outcome. There is no direct evidence that the Oedipus complex was revealed or relieved by psychoanalysis.

Even Hans' fear of horses and his dreams are presumed in psychoanalysis to be symbolic of something else, but the basic idea in scientific inquiry is to let the facts speak for themselves. If we adopt the evidence-demanding attitude of science, support for the Oedipus hypothesis in these instances comes too much from Freud, too little from the facts of the case.

In response to all this Freud might say that it was his fate merely to discover the obvious. He could not understand why people were so blind to the role of sexual factors at all stages of life. In reply to objections regarding Little Hans he pointed to the many details that other explanations ignored. How does one explain the importance of the blinders and muzzle except according to the Oedipus hypothesis? How does one explain the giraffe dream? The sheep and bathroom dreams? And what was the significance of the heavily loaded horse?

For what reason did the boy later have his father marry his grandmother in fantasy? Why was he particularly fearful at bedtime? How does one explain the sudden concern about horse bites when the warning in Gmunden preceded the phobic reaction by at least eighteen months? It was Freud's view, held by others as well, that these dreams, fantasies, and other concerns were called into service later, in response to the recently awakened Oedipus impulse.

A good hypothesis organizes our knowledge about a problem and explains the facts, and it was Freud's contention that no other approach did as well as psychoanalysis in this instance. A good hypothesis stimulates research and provides direction, and Freud's work certainly has been fruitful in this way. But a good hypothesis is also testable and much of Freud's thinking, developed in a clinical domain, has not readily lent itself to further examination.

The best we can say now about the Oedipus-Electra hypothesis in the case of Little Hans, and also as a universal phenomenon, is that some findings have been made and some important parts of the puzzle are still missing. The hypothesis at this time remains uncertain.

Freud's interests in this hypothesis were received with considerable skepticism in his own time and hotly debated in academic circles. By the public they were considered scandalous, a case for the police, to be discussed only in hushed tones.

Cafe Conversations. *Particularly in the sidewalk society of the Ringstrasse, early psychoanalysis was regarded as subversive.*

But times have changed and the morality of this work is no longer a significant issue, even at the Ringstrasse. Of special importance instead is the possibly confounding role of Freud's work with Little Hans' father. In his long walks and talks with the boy, as cotherapist, the father made several remarks that perhaps fostered the outcome Freud sought, consistent with psychoanalysis.

"Daddy, you *are* lovely! You're so white," the boy said one day. "Like a white horse," the father replied, reflecting the psychoanalytic perspective.

178

On a return trip from the country the man was talking with his son about the size and color of the horse that had fallen. He asked, "When the horse fell down, did you think of your daddy?"

"Perhaps. Yes. It's possible," answered Little Hans.

The father, an early, enthusiastic supporter of psychoanalysis, was determined to assist in whatever way possible. Maybe at times he helped too much.

This part of the Hans' legacy, the story of the boy, illustrates the clinical method, aspects of psychoanalysis, and something more. Like the tale of the horse, it also shows the potential for investigator influences in research. Just as Freud unwittingly encouraged recall of childhood seductions in his early patients, just as the father perhaps encouraged statements supporting the Oedipus hypothesis in his son, so any clinician may unknowingly encourage research subjects to respond in a way that is favorable to a certain interpretation of the case.

This possibility—a research outcome attributable more to the hopes of the investigator than to the treatment provided—is the central theme of the Hans legacy. It is the basic point of contact between the two tales, beyond the coincidence of names, dates, locations, horses, and use of the scientific process. These cases are regarded today not only as early models of experimental and clinical methods in psychology but also as classic demonstrations of possible unwanted influences on the part of the investigator in the research process.

19 Berggasse

Part Five
The Debt Repaid

CHAPTER THIRTEEN

Reports

Pfungst, for all his efforts, is a forgotten man. He looked into the mind of a horse and found less than was expected. Freud, for all his efforts, is a famous man. He looked into the human mind and found more than was expected.

Pfungst discovered unconscious signals that guided the clever horse. Freud allegedly found unconscious motivation that disrupted the little boy. Pfungst was a patient man who toiled mostly in obscurity. Freud was dramatic and ambitious, and today 19 Berggasse is a museum of international interest. But after their studies of the horse and the boy, Pfungst and Freud each had one task remaining.

They owed the scientific community a *research report,* which is a complete account of what took place in an investigation. This report makes verification possible; it provides for continuity and growth in the field; and it preserves knowledge for future generations, to be used or neglected as they see fit. Without these reports there would be essentially no psychology, certainly no Hans legacy, and much less understanding of the scientific process.

Pfungst's report showed the usual experimental format. After a brief history of the problem and statement of purpose there appeared a description of the *method,* which is a major section in any experimental report. It describes how the research was accomplished, with emphasis on the subjects, apparatus, and procedure.

Clever Hans was Pfungst's only animal subject, and he described the horse in detail, including his breeding, background, and training, giving special attention to the instruction received from Mr. von Osten. The chief apparatus included the blinder, screen, pins, placards, and colored clothes, all kept in the shed in the courtyard. Most important were the various experimental and control procedures, also reported in precise fashion.

Many human subjects were used in the verification studies, as is common today, and Pfungst stated the age, sex, and occupation of these people. He measured their head movements by adapting an instrument for detecting hand tremors, and he employed diverse signals and "blank" trials, all as a check on the accuracy of his work.

After the method, the *results* appear. The basic findings are presented as objectively as possible, usually in numerical form. The advantage of numerical data is that they can be highly precise; differences among conditions and outcomes can be measured and specified. According to Sir Francis Galton, cousin of Charles Darwin and father of modern statistical methods: "Until the phenomena of any branch of knowledge have been submitted to measurement . . . that branch cannot assume the dignity of a science." Some researchers might not concur with Galton, but most would agree that *quantification,* in which phenomena are counted or measured in some way, has been indispensible in modern science.

Pfungst quantified his findings by observing the rates of success, types of errors, and reaction times, and these data were displayed in tables and graphs. Hans' errors in selecting a colored cloth, for example, could not be explained on the basis of one color being mistaken for another, for when a brown cloth was requested, the animal at various times selected the orange, the green, or the yellow. There was no systematic color error in the horse's mistake.

Eventually it was discovered that the answer lay in the position

Color Test. *Hans selected the fourth card by touching it with his nose.*

of the cloth. When Hans erred, he usually selected the adjacent cloth, missing the signal slightly. Rarely did he miss by a wide margin. With Mr. von Osten as questioner, 73% of Hans' errors occurred when he chose a color next to the target, and only two percent involved a color four or five places away. Comparable results were obtained with Pfungst as questioner:

Hans' Errors With Colors

Questioner	Number of Places in Error				
	1	2	3	4	5
von Osten	73%	21%	4%	1%	1%
Pfungst	68%	20%	11%	1%	0%

Quantifying his observations in these ways, Pfungst came upon several additional questions that he proceeded to test, one of which was this ad hoc hypothesis: only the person with the greatest control over the animal at any given time influenced the result. Spectators, whether or not they knew the answer, had little effect, and several people concentrating simultaneously made no difference either.

To test this hypothesis Pfungst competed with his professor. He and Stumpf stood beside the horse, each thinking of a different number. Stumpf concentrated on five and Pfungst on eight, and Hans answered with the larger number. Stumpf concentrated on seven and Pfungst thought of four, and the horse answered with the smaller one. When Stumpf had six in mind and Pfungst did not concentrate upon any number, Hans tapped thirty-five, apparently awaiting Pfungst's signal. Finally Pfungst left the area, Stumpf again thought of six, and Clever Hans tapped that number. But when Pfungst returned, all Stumpf's attempts failed. Like von Osten before him, Pfungst's long efforts with the horse meant that he had become its master.

At the end of the report there is a *discussion* of the entire project, in which the investigator presents an overall interpretation of the research. Here the facts do not stand alone; rather, the case is argued. With greater freedom than earlier, the investigator explains the significance of the work, its successes and limitations, and its place with respect to theory and other research. In science one must make discoveries and then make something of these discoveries.

Oskar Pfungst at this point considered the question of telepathy. The "mind reader," he said, may be guided partly by involuntary expressive movements of the head and other body parts. He cited an American neuropathologist who believed that this sensitivity should be called "muscle-reading" or "body-reading," so dependent is it upon the perception of extremely minute muscular movements. The potential in daily life is enormous, Pfungst claimed, and he suggested further inquiry into this topic.

This report should have been quite final as far as the horse was concerned, but among the public there was still talk of the occult.

One writer, ever loyal to telepathy, insisted that "the horse receives the thought-waves which radiate from the brain of his master." Several days later a reporter for the *Berliner Tageblatt* claimed that the horse did not use visual cues: "Tests have been made upon Hans with blinders over his eyes and it is to be noted that, in spite of these, he still responds correctly." These reactions show that

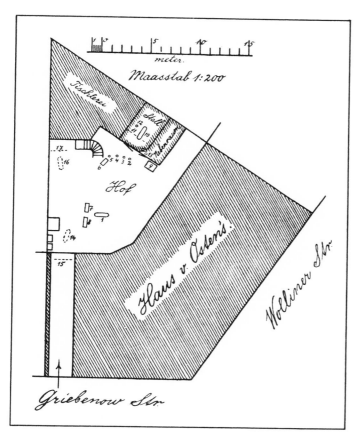

Diagram of the Griebenow Courtyard. *Later investigators made a drawing of the "Hof" or courtyard, showing the position of the horse (1), their own positions (2–5), their writing table (6), Hans' work tables (7,8), and even the "Tischlerei," which is the Piehl's carpentry shop.*

while some people do research, others are left to interpret the results, and sometimes they make of them whatever they wish.

Pfungst's report deviated only in minor ways from the modern viewpoint of behaviorism. Every report must be written from some perspective and his, despite its early appearance in the history of psychology, reflected this outlook. He studied only readily observable events, which is the basic method of behaviorism, and eventually he discovered that two stimulus-response connections accounted for most of the horse's performance: tilting the head forward was the signal to begin tapping, and tilting it backward was the signal to stop. Altogether his work was highly objective, and in this respect it was certainly consistent with early behaviorism.

He subtitled this report "Der Kluge Hans," meaning simply

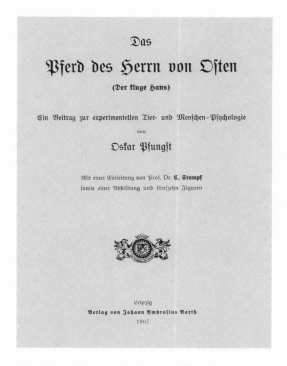

Der Kluge Hans. *Published in 1907 with a lengthy introduction by Carl Stumpf, the book was translated into English four years later.*

188

"Clever Hans," and it contained a photograph of the horse and several tables and graphs. Widely disseminated at the time, down through the history of psychology it has been regarded as a classic study in the experimental method.

So much for Hans-the-horse. Freud's report of the boy was also *Clinical Report* widely disseminated and had the same basic purpose: to provide a record for later investigators. It was first presented in lectures in the United States in 1909, where Freud had come to speak on psychoanalysis. In these talks, delivered in German without notes, he made frequent reference to "Der Kleine Hans," meaning the case of "Little Hans." Later that same year, to provide a more enduring record, he published the study in his journal.

JAHRBUCH

FÜR

PSYCHOANALYTISCHE und PSYCHOPATHOLOGISCHE FORSCHUNGEN.

HERAUSGEGEBEN VON

PROF. DR. E. BLEULER und PROF. DR. S. FREUD
IN ZÜRICH,   IN WIEN.

REDIGIERT VON

DR. C. G. JUNG,
PRIVATDOZENTEN DER PSYCHIATRIE IN ZÜRICH.

I. BAND.

I. HÄLFTE.

LEIPZIG und WIEN.
FRANZ DEUTICKE.
1909.

Der Kleine Hans. *The case of Little Hans was the first article in the new yearbook; it was titled "Analysis of a Phobia in a Five-Year-Old Boy."*

On his arrival in America Freud was received with little enthusiasm and less understanding. When his ship docked, a New York newspaper announced the visit of Professor Freund. City newspapers in Worcester, Massachusetts, where he spoke at anniversary celebrations at Clark University, gave more attention to a man who talked about school sanitation, and here he was referred to as Dr. Edmund Freud. His field of endeavor was called phychology, his work was that of a therapeutist, and he was said to be concerned with conscientiousness. Nevertheless, in the course of five lectures he made an increasingly favorable impression. Newspaper coverage broadened and became more accurate as the week progressed.

One Boston correspondent later described Freud as a person of great refinement, of intellect, and of many-sided education. Another reported that his views "attracted the eager interest and in many cases won the adherence of the scientists assembled." Still another concluded that his works were "nothing short of epoch-making. The 'Story of Little Hans' will probably ever remain a unique and model study of a child's soul."

Freud's discussion of Little Hans was suitably disguised, for the clinical report is a private matter, concerning the adjustment of a particular individual. It generally remains confidential, except when a new method has been employed or a new finding is uncovered. Then it is published with pseudonyms and used to stimulate further developments in the field.

Public or confidential, the clinical report follows a common form, beginning with a brief statement of the patient's problem, called the *presenting problem*. Little Hans' presenting problem was that he refused to go out of his house, apparently because he was afraid of horses.

This information is followed by a much longer section known as the *background* of the case. Taken from the case history, it summarizes the person's life, highlighting incidents relevant to the presenting problem. Little Hans was especially fearful of venturing into Lower Viaduct Street, and the father decided that this information should be communicated fully to Freud. He sent a diagram of the neighborhood along with his weekly notes on Hans' condition. Across the street from Hans' house was a courtyard with a warehouse, and through the railing of the fence horse-drawn carts could be seen at the loading dock. Freud included this information as part of the case.

The father noticed that Little Hans was particularly frightened when the carts went through the gates, at which time they had to turn a corner. The boy explained: "I'm afraid the horses will fall down when the cart turns." But he was also afraid whenever a waiting cart started to drive away.

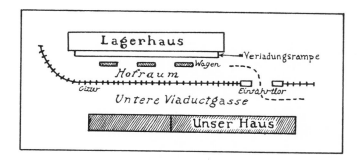

Diagram of Lower Viaduct Street. *Three wagons are shown at the "Verla-dungsrampe," or loading dock, before passing through the gates. "Unser Haus" designates "our house."*

The next section, *diagnosis*, indicates the clinician's understanding of the problem. It presents the results of interviews and psychological testing, concluding with a diagnostic statement. For Little Hans, the diagnosis was: "Neurosis, phobic reaction."

Use of the term "neurosis" is declining today, partly in resistance to the long-standing dominance of psychoanalysis in clinical psychology. Some psychologists feel that it gives too much emphasis to psychoanalysis and to the concept of unconscious motivation; others simply prefer terms with more precise definitions. Persons in Little Hans' condition are now diagnosed as "phobic disorder" or "simple phobia," a change intended to recognize the diverse views and need for precision in contemporary clinical psychology.

Finally, there is a section on *recommendations* for treatment, in which a remedial program is described. Here tentative dates are set to reassess the individual and to determine what changes in therapy, if any, may be appropriate in the future.

Freud's reports did not follow this format closely; they reflected instead a freer style for which he was renowned. These writings brought him the Goethe Prize for literature, as well as other fame, but his fundamental aim was to promote the psychoanalytic viewpoint. All his reports should be understood in this context, as stressing unconscious processes and thereby furthering the cause of psychoanalysis.

After the lectures, Freud went on a brief vacation in the Adirondack Mountains, and then he returned to 19 Berggasse. Having

Freud in the Wilds. *Before leaving Vienna on his difficult mission in America, Freud announced that he had two purposes: to speak on psychoanalysis and to see a wild porcupine, the latter goal being a fortification against possible failure of the lectures. Staying in this cabin, he found a lifeless carcass of the beast, which he poked with his walking stick, and then he returned to Vienna, having achieved his double goal.*

accomplished his purpose of presenting psychoanalysis on another continent, he was in a position to enjoy the beginnings of an international reputation.

At times, a certain viewpoint or perspective in psychology be- *Systems of* comes a way of approaching the whole field. Then it is called a *Psychology* *system of psychology*, for it provides a general framework; it serves as a means for deciding which facts are to be obtained and how they are to be ordered in relation to one another, since all psychological phenomena cannot be considered in any given instance. Both psychoanalysis and behaviorism organize psychological knowledge in this way, and they have been regarded as systems of psychology. But they are contrasting approaches, as illustrated by two important differences between them, one of which concerns the time factor.

The emphasis in behaviorism is on the present. Oskar Pfungst preceded the behavioristic movement, but he showed this spirit of inquiry, focusing on Clever Hans' behavior as it was revealed in the courtyard. Psychoanalysis is largely concerned with the past. Sigmund Freud studied Little Hans' behavior at the time of the interview—but only because Little Hans was a young child. Had Hans been an adolescent or adult, Freud would have encouraged him to reconsider his earlier years. Psychoanalysis delves into the past; behaviorism is concerned with the present.

But the two systems are not completely distinct in this regard. Psychoanalysis recognizes that current circumstances can evoke memories of the past. And behaviorism recognizes that certain early reinforcers can maintain present behavior. The matter is one of emphasis.

The second difference concerns the response, for behaviorism focuses almost exclusively on overt actions. Oskar Pfungst studied the tapping of the horse and the movements of his questioners, both of which were overt. They could be observed directly. Traditional behaviorism gives little attention to internal processes, which must be inferred. This insistence on external events and readily measurable phenomena is sometimes called the "psychology of the empty organism."

Psychoanalysis instead concentrates upon events inside the individual. Inevitably these have some physiological bases, as when adrenalin is secreted in emotion and neural activities occur in thinking, but the emphasis is on mental processes and feelings, which cannot be observed directly in someone else. Freud was particularly interested in dreams, which are also internal events, and these have proven similarly elusive in research. Hence, psychoanalysis is sometimes said to be imprecise and subjective.

On this basis the reader may wonder which system provides the better understanding of human functioning. This decision is a matter of personal preference. Does one prefer a lower-level but more parsimonious explanation, as in behaviorism, or a higher-level, more speculative one, as in psychoanalysis?

The decision also depends upon the problem under examination. Behaviorism might have been tried with the boy, using various techniques to encourage him to venture into the street, but psychoanalysis of the horse would have been a dubious enterprise.

For such reasons many modern psychologists do not commit themselves fully to either system. They find parts of one or another system useful, and many remain asystematic, proceeding along an independent path. The emphasis on single systems of psychology, which emerged at the time of Clever and Little Hans, has diminished in recent years. There are instead several contemporary definitions of behaviorism and many varieties of psychoanalysis. These and other systems now represent rather generalized ways of viewing the world, and they are adopted by psychologists who, as a result of training and experience, have interests in different psychological questions. As diverse approaches to complex problems, they give the field a hybrid vigor.

CHAPTER FOURTEEN

Recognition

After his work with Clever Hans, Oskar Pfungst for a short time became a celebrity in Berlin. He wrote articles for the press, gave lectures, and then added to his reputation by investigating horses from the circus, the cavalry, and riding schools.

To a surprised circus trainer he demonstrated horses' inability to understand spoken words. He showed that an animal at a full gallop could not be brought to a trot or a walk by any call whatsoever, friendly or hostile. The horse simply made no response. The slightest gesture with the reins, the whip, or the boots brought forth a reaction but the verbal signal was fruitless.

Army officers agreed that cavalry horses understood spoken commands, and one pointed out that they surpassed the new recruits in this regard. But Pfungst was doubtful. He prevailed on Captain von Lucanus to lend him three soldiers and their mounts, selecting the latter as the most intelligent in the regiment. Then he again demonstrated his point.

When walking or trotting, the animals neither increased nor decreased their pace at the sound of a spoken word. The type of command made no difference. Almost any signal prompted a standing horse to begin moving, including a bugle call and even the word "retreat." The horses could tell which commands were directed at them and which at the riders, however, for when Pfungst called "Lances down!" they did not move. After some further work, he decided that horses were not particularly interested in sounds anyway.

Follow-up Studies   In these studies Pfungst moved from one question to another to still another. Discovering the unconscious signals with Clever Hans, he then looked at his own ability to detect them, at others' ability, at the state required to transmit them, at individual differences in sending and receiving them, at circus and cavalry horses, and so forth, each time pushing beyond the original problem and even beyond his verification of that answer.

In this respect scientists are like the ancient Greek warrior challenging the seven-headed Hydra. For every head that is cut off, two more grow in its place, and the dragon becomes all the more awesome. For each question that is answered in science, others arise to replace it. The price of getting a head is that there is more work to be done. Hercules solved the problem by cauterizing each wound, but the solution in science is not so readily encountered.

This continuous process is part of all successful research. The purpose is to provide more and more precise answers to scientific questions. These subsequent efforts are called *follow-up studies*, for some earlier question is pursued in further detail or an associated question is considered. We shall consider some of these outcomes in the work of Pfungst and Freud, but we turn first to the trainer and his horse and then to the boy and his father, examining the outcomes for them.

For some months after Pfungst's report the volatile schoolmaster lamented the injustice of it all. "I would consider it impossible that something like this would happen," exclaimed Mr. von Osten, suggesting that his case would have gone better in other countries. "I am very sorry about the matter but I want to give my Hans away to someplace in America because he is underestimated in his own

196

fatherland." The old man was unpredictable, however, and later wanted to punish his learned horse.

One Sunday morning he appeared in the courtyard wearing a mask, planning to test the horse while hiding his facial expression. But with this encumbrance he found it difficult to know whether the horse had answered properly. Everything happened too quickly.

The old man had toiled in a careful, patient fashion, and he truly believed that he had trained Clever Hans to think for himself. On hearing Pfungst's findings he became grief stricken, prohibited further studies of the horse by anyone else, and decided that the scientists had unintentionally trained Hans to respond to their signals. Now it was his task to break him of the habit. He persisted

Wilhelm von Osten. *After Pfungst's investigation he blamed his once-famous steed for his own misfortune, and he decided that Clever Hans should spend the rest of his life carrying bricks and mortar in repentance.*

for a while and finally gave over the use of his beloved creature to a wealthy jeweler from Elberfeld. Regardless of Pfungst's evidence, he thought his honesty was in doubt.

His tireless and expert pedagogy had brought forth no indication of abstract thinking in the horse. It constitutes instead the clearest support for a broad gulf between human and animal thinking. It is, in the final analysis, a most convincing demonstration that conceptual thought is not to be found in lower animals, at least up to the evolutionary level of hooved mammals. This unintentional outcome is Mr. von Osten's most important contribution to science, besides prompting inquiry of the signals in the first place.

These signals, we now know, were not an isolated instance in Berlin at the turn of the century but rather are a general law of verbal behavior, one which seems to apply to all normal persons of Western Europe and North America. In English it appears that every native speaker raises the head almost imperceptibly after saying something to which a reply is expected. Each of us apparently learns this postural request for a response in our earliest years of language acquisition, just as we learn to raise the pitch of our voice with a question, lower it with a statement, and make no change when we intend to continue speaking. In acquiring a language, we learn the necessary sounds and also the accompanying movements.

But these findings could offer no consolation to the poor old man, who died not long thereafter, apparently of a broken heart. He expired in mid-afternoon on June 29, 1909 after a disease of the liver had kept him bedridden for six months. The neighbors at his grave all talked of the hermit-like quality of this solitary man. Industrious and ingenious, he remained completely alone, his head crammed full of ideas from Darwin, Gall, and even Annie Sullivan, teacher of Helen Keller. He was, however, absolutely unschooled in the most basic requirements of the scientific method, and certainly he had been frustrated in the great goal of his life.

So much for the horse, his cantankerous master, and their brief hour of triumph, long since forgotten by most everyone. More important as a follow-up study in psychology, and for the Hans legacy in particular, is the outcome for the fearful little boy in Vienna. He was the subject of a controversial new approach to the study of personality. How did things turn out for him, his parents, and Freud's other patients?

Of all Freud's cases, incidentally, the most famous is probably Anna O., who later became a founder of social work in Germany, prompting that government to issue a commemorative stamp in her honor. She also participated in early feminist movements but apparently could not save herself from a recurrence of mental illness.

Anna O. *Dressed here in riding habit, years before her death she wrote several obituaries for herself, all of them humorous.*

Whatever credit may go to Freud for whatever gains she made in personal adjustment must be tempered by this information. In any case, her real name is Bertha Pappenhaim and she is remembered in the present context for her early role in the "talking cure".

As for Little Hans, Freud never saw him again after that moment on his doorstep, but he did receive a second-hand report seven years later. When Freud was in Frankfurt in 1930 to accept the Goethe prize of several thousand dollars, the city counselor paid him a visit before the ceremony. Much to Freud's surprise, the man was accompanied by his own wife and also the bride of Little Hans, many miles from home. She reported that her husband, now aged twenty-six, was well adjusted and in normal health. By his marriage, according to the psychoanalytic view, Hans had further resolved the Oedipus problem—in the same way as his father before him. One wonders, to carry the Freudian hypothesis to its limit, to what extent Hans' wife resembled his mother.

But the wife's report of normalcy omits many later decades of Hans' life, during which he could have entered psychoanalysis again, for many people come to this treatment for the first time at age forty or later. All of which emphasizes again the vital role of follow-up studies in psychological research.

The final outcome for Hans' father was a bit unexpected. He remained a devotee of Sigmund Freud but probably not of his own wife. They were divorced a few years after Hans' analysis, a relatively rare occurrence at that time. And that is all we know of this mustached, bespectacled Viennese physician who liked to give horsey rides to his son and assisted in the first psychoanalysis of a child, in rather unorthodox fashion.

The man is well remembered, however, for the interview with Freud. That moment in the consulting room at 19 Berggasse was in some respects the beginning of modern psychotherapy with children. The term *psychotherapy* refers to a variety of remedial techniques with emphasis on a conversation between patient and therapist, which is what happened when Freud met with Little Hans. Physicians and specialists in education of course had worked with children previously in this fashion, but Freud, with his early and strong interest in the interview, is generally considered the founder of modern psychotherapy, at least in Europe. And today there are many variations on his original method.

As these cases of Little Hans and Clever Hans both demonstrate, science is basically unfinished business; it is a constant progression in the light of new evidence. Science involves a search, re-search, and further research, and the story is never fully told. No matter how complete it may seem at the moment, any given finding, method, or interpretation is just a a step along the path, to be re-examined or extended at some future date.

Pfungst, when finished with the cavalry horses, returned to his laboratory, and then he was diverted by another public call, this time to study a clever animal called Don the Talking Dog, renowned for ability in human conversation. Don could introduce himself and then discuss people and events around him. When it was noisy he was asked "Was bittest du dir aus?" That question meant "What do you want?" The dog answered "Ruhe," meaning quiet. His accomplishments were widely publicized, even in America. *Investigator Styles*

Pfungst handled the case readily. To find out whether the animal really understood people, he asked the questions in one order, then another, and then another. The result was that Don sometimes said his name was "quiet" and sometimes he wanted "hunger." He simply did not know what he was saying; his answers were nonsense.

To find out whether people really understood the animal, Pfungst controlled the context of the dog's speech, making phonograph records and playing them for listeners who had no idea of the situation in which the animal was barking. They could seldom distinguish his "Ruhe" from his "Kuchen" and were just as likely to interpret it as "Huhn," meaning chicken, or even "Hallelujah." Don's "words" were all in the listener's ears; his audience made sense out of his meaningless noises.

Oskar Pfungst worked with characteristic thoroughness, accumulating as much data as possible in support or refutation of any hypothesis. In an organized, careful way, he was content with small advances on modest problems. This approach is described as *systematic research,* for the investigator proceeds in a meticulous, step-by-step fashion, making as few guesses as possible, recognizing that scientific progress is usually slow.

Pfungst certainly was systematic with the horse, even after he discovered the basic answer, and he showed this same persistence again in the laboratory. Would most people send the signals? Could he learn to detect them? Do people differ in sending and receiving them? Pfungst was invariably thorough, analytic, and patient.

Pfungst and the Horse. Oskar Pfungst *gained a transient reputation for his work with* Clever Hans. *Like Wilhelm von Osten, shown at the left, he was highly systematic.*

Sigmund Freud was different. He was thorough, especially in his early laboratory studies and later scrutiny of minor daily incidents, but he was not always patient. The impulse to *speculative research* was always within him; he considered himself not an investigator but an adventurer, a "conquistador" in the world of knowledge. He began his career by studying animal tissue through a microscope and concluded it by studying human beings through their dreams. In all his research one can see a great eagerness to move ahead into new territory. Behind a concern for proof there was always a longing to give free rein to imagination.

Sigmund and Martha. *His dramatic style was reflected in their courtship, during which he wrote his fiancee some 500 letters, often including one red rose and some remark in a foreign tongue.*

In speculative research, as shown in much of Freud's work, the investigator is more inclined to take a chance, to guess at results, and to move ahead rapidly. Freud was bored whenever he learned nothing new or had no novel ideas, regardless of the patient's progress. He proposed his theory of personality after working with but a few cases, and a friend described his intellect at the time as "soaring like a hawk." His great contribution to psychology, like that of other great figures in the history of science, is to be found in his basic reformulation of the problem. His thinking was revolutionary; he proposed an entirely new mode of inquiry and a new way of looking at the results.

Freud is recognized today for at least four major contributions to contemporary psychology. He was the leading spokesman for the clinical method, for research on dreams, for redirecting Western thought on the origins of adult personality, and especially for the novel, challenging concept of unconscious motivation. No student of human behavior can afford to overlook the impact of his thinking on the manners and morals of contemporary Western society.

But sometimes nothing is found in speculative research. The effort seems foolish or wasted. It is scorned for the moment and then forgotten, as happened to Blondlot with N rays and Freud in several instances. Freud moved so rapidly with the seduction hypothesis that he apparently misconstrued the data, and later he was so confident about his work on dreams that he overstated the case. A more modest inscription on that tablet in the Hotel Bellevue might read: "Here some provocative thoughts on the early study of dreams were first expressed by Dr. Sigmund Freud."

There is of course a constant interplay between speculative and systematic research. One more impetuous, the other more cautious, they often produce complementary research outcomes. Newton was only partly correct when he spoke of standing on the shoulders of giants. Advancements in science depend upon the great leap forward, often the unmistakable achievement of some speculative researcher, and upon the countless lesser contributions of those more systematically inclined, who gather the bits and facts from which some bold, new step is taken. Their earlier work provides a springboard for the sudden leap forward, and their later work is devoted to verifying, disproving, or reformulating the new perspective.

204

Freud's name in psychology is thus far greater than Pfungst's. Remembered for his successes, commonly forgotten for his failures, he has become an eponym, which is a likely outcome for the successful, speculative researcher. An *eponym* is someone so prominently connected with a certain event or way of thinking that his name becomes the name for that event. Such instances in this story include Darwinian biology, Skinnerian psychology, Morgan's canon, and even the Ferris wheel. In each case someone's name has become synonymous with a new perspective, principle, or other invention.

We now speak of Freudian theory, Freudian psychology, and the *Freudian slip*, referring to some small mistake in speech, writing, or memory that presumably reflects unconscious motivation. In the Freudian slip it is assumed that some early, repressed conflict breaks the barrier of repression and appears as symbolic behavior. Freud himself, for example, mistakenly wrote about Hannibal's father as his half-brother, and Little Hans forgot that his father wore glasses, both errors suggesting an Oedipus complex not yet resolved. There is also the Freudian approach to dreams, and in a more general sense, someone is said to be Freudian whenever that person's view of human behavior gives particular emphasis to unconscious processes. Freud is one of the most eponymous figures of the twentieth century.

But is this use of eponyms justified? One view of history favors such name-giving, emphasizing that "the person makes the times." The tremendous revolutions in human thinking that followed the work of Copernicus, Darwin, and Freud are offered as testimony. These individuals, often speculatively inclined, influenced in a very fundamental way our whole conception of human life.

From another perspective the crucial ingredient is the context within which the individual works. When science has reached a given stage, certain discoveries are almost inevitable, and hence "the times make the person." Eponyms owe much to the less timely, less boldly inclined multitudes who preceded them—to the Oskar Pfungsts and Hugo Munsterbergs—who have laid the essential groundwork, developing the knowledge that antedates any great discovery. These people, of lesser significance individually than the name-givers, collectively make a greater contribution to science.

Someone else in the sixteenth century, according to this view, would have developed the heliocentric concept of astronomy had Copernicus not come upon the scene. He lived at the right time to make this discovery, which places the sun at the center of our universe. Someone else, in fact, did make Darwin's contribution at just his moment in history. A countryman, Alfred R. Wallace, independently produced the theory of evolution and sent a copy of his manuscript to Darwin, who then hastened to publish his own.

Freud's approach to psychology, more than any other in this field, is the product of one mind. But it too was foreshadowed in small ways by others, most notably the German philosopher Arthur Schopenhauer and to a lesser extent the French neurologist Jean Martin Charcot.

Why then do eponyms arise? Scientists typically do not suggest this use of their names, and when eponyms are compared with their forgotten brethren, it is clear that the latter have anticipated almost all the great advancements in the history of science.

It seems instead that the name is bestowed by an admiring public, much in need of simplicity and heroes. It simplifies things considerably to believe in great scientists, for an eponym brings within our understanding a great sweep of history, a whole era of investigations by countless systematic researchers. It is also satisfying to think of great scientists. Eponyms give humankind, always in need of leaders, some of its most enduring heroes. Perhaps the distortion of fact is worth this gain in human welfare.

The name of Oskar Pfungst is thus lost from our lips, except in a tale such as this one, for the horse proved to be an ordinary creature and Pfungst a dedicated scientist. But Clever Hans' susceptibility to cueing was so marked, the case so controversial, and the finding so important to psychology that instead the animal became an eponym. The horse displaced the scientist and today, whenever unintentional cues are given to anyone, this outcome is known as the *Clever Hans phenomenon*. Within and outside psychology the unwitting transmission of subtle signals is referred to in this way. The otherwise dull horse and the founder of psychoanalysis are thus remembered together, as name-givers in early psychology.

CHAPTER FIFTEEN

Human Inquiry

It seemed that Clever Hans did a complex thing, but Pfungst's explanation was simple. It seemed that Little Hans did a simple thing, but Freud's explanation was complex. Oskar Pfungst and Sigmund Freud, each in his own way, demonstrated the scientific process in psychology.

Pfungst and Freud, in the course of this research, also demonstrated something else. They showed that the story of science inevitably bears a human imprint. They showed that the scientific process, in a most fundamental sense, is influenced by the investigator's hopes and expectations for a fruitful outcome. The investigator's personality is an inevitable part of the research process.

The human element, our final concern in this story of science, appears in diverse forms known as investigator effects. In all instances of *investigator effects* the scientist unknowingly influences the research outcome, producing a finding that is not exclusively attributable to the events under examination. It is instead due partly to the fact that the investigator, in studying a certain condition, has therein altered it in some way.

After Pfungst's findings had been made public, the volatile horse trainer became even more unpredictable. Despondent one moment, forgetful the next, on several occasions he explained Pfungst's results in this way. He said that the horse's behavior had been altered by the scientists who came to investigate him. In their work they had interfered with Clever Hans' normal performance, unintentionally training the animal to respond to the questioner's head movements, and now it was his task to break the horse of this unfortunate habit.

Mr. von Osten was wrong about what Pfungst had done, but he was right about the scientific process. The act of studying something can alter it. An investigator does not observe some event "out there" in nature, completely independent of all external influence, but rather studies that event as it is exposed to the process of human inquiry. The investigator examines a system of which he or she is a part and thereby influences it in some way.

Errors of Observation Our story began with the feats of the marvelous Berlin horse, scrutinized for tricks of all sorts. But no signals were identified, not even by thirteen persons from every walk of life deemed relevant to the animal's success. One popular explanation involved N rays, an embarrassing episode in the annals of science but one certainly worth remembering, for it stands as an object lesson in scientific inquiry.

Once the new concept took hold, one can only marvel at the rapidity with which N rays invaded the halls of science. After the first announcement in the proceedings of the French Academy of Sciences in March 1903, N rays were observed in all parts of the world. Professor Blondlot encountered only three or four people who could not detect the new radiation, and Blondlot himself could observe and even measure it, calculating the specific wavelengths with an accuracy greater than 96%. It took a very strong belief in these nonexistent rays to see them in the first place, to measure them by different methods, and then to arrive at such compatible values.

In Vienna at this time it was generally believed that infantile sexuality was impossible, that children were completely innocent of sexual interests, and that only a depraved man like Sigmund Freud could think of such things. Scientists and laypersons, rather than manufacturing false evidence, as in Nancy, were ignoring the true evidence. As Freud said, it was his fate merely to discover the obvious, which "every nursemaid knows."

There is an old saying that describes these circumstances: "We see not what lies in front of our eyes but what is behind them." We see what we expect to see, hope to see, or are used to seeing, and thus the first category of investigator effects is known as *errors of observation.* These occur simply because every investigator has some personal limitation or some tendency to expect a particular outcome. In obtaining evidence, even the most conscientious researcher can make mistakes of this sort.

Just as observation can be distorted by human inclination, so can human reason. Scientists can make errors in the assessment of evidence, in deciding what to make of an observation, and these investigator effects are known as *errors of interpretation.* *Errors of Interpretation*

Mr. von Osten readily demonstrated this error, chiefly because he was so fond of the horse and his own apparent accomplishment. When Mr. von Osten gave the command "left," Hans gestured to his own right, and his owner interpreted this behavior as demonstrating the horse's capacity to take another person's viewpoint. When the horse missed a simple question, Mr. von Osten declared that Clever Hans had a will of his own. When Hans answered a difficult problem with greatest ease, his master reasoned that it was because he was in a better mood. Almost everything his beloved animal did was interpreted by Mr. von Osten and some others as a sign of extraordinary ability, so eagerly did they project their own hopes onto the animal.

It was necessary, Pfungst noted, to record these alleged signs of the horse's genius, so seriously were they misinterpreted by soberminded people. According to many, Hans' intelligent eyes, his high forehead, and especially the carriage of his head showed "a real thought process going on inside." If Hans turned toward a visitor who made some comment in praise of his performance, it was a sign of his appreciation, and if the horse failed a test, it was because of his stubbornness. Like the folk who witnessed Don the Talking Dog, Mr. von Osten and others made sense even out of Hans' mistakes.

In a different setting, Freud listened to patients as they lay on a couch and whispered their innermost thoughts, and it seems that he too made errors of interpretation. Eagerly searching for the origins of neurotic disorders, he decided that the evidence for child-hood seduction was irrefutable. He even regarded the resistance in his patients' reports as further support that the episodes could not have been invented in the first place. Otherwise, why should the patients wish to deny them? The conviction with which he interpreted these elaborate fantasies as absolute proof of seduction is impressive.

Later Freud decided that the origins of neurosis lay instead in the infantile wish for incestuous relations, and critics have said that here too he misinterpreted the data. According to this view, the evidence does not allow even this interpretation; Freud simply chose to regard his patients' thoughts in these terms.

There is also some question about Freud's interpretation of Sophocles' story of Laius and Oedipus, father and son. Rather than an Oedipus complex, it might depict a "Laius complex," in which the father receives a just punishment for abandoning his child. Oedipus had no special desire to kill his father and marry his mother, for he did not even know them. Did Freud prefer to lay the blame on the son, including himself, rather than implicate his father?

Criticisms of Freudian theory point to possible errors of interpretation in his thinking, but they also depict the compelling quality of his work. For example, Freud was rebuked at the turn of the century for adopting the seduction hypothesis, and eighty years later he was denounced again, this time for discarding it in the first place. His work has prompted endless scrutiny, yet his perspective on psychology has been an unprecedented source of interest for scholars of all sorts, so much so that it is difficult to imagine twentieth-century thought without his ideas.

Errors Through Cueing   Apart from errors of observation and interpretation, and except for cases of intentional fraud, there is a third category of investiga-

210

tor effects, found only in the behavioral sciences. These errors arise in the relationship between the investigator and the subjects, and they act directly on the subjects' behavior. They are called *errors through cueing* because the investigator, by some subtle and unexpected means, informs the subject of the behavior that is sought and anticipated.

This unwitting transmission of information, brought to public attention in the case of Clever Hans, is certainly the most important outcome of this research. Pfungst found it in the courtyard and the laboratory, and he demonstrated the conditions under which it occurs, emphasizing that the requisite state is not a passive expectation. Rather, the person must possess an air of quiet authority or intense concentration, making the expectancy all the more forceful and probable. It happened with Pfungst's laboratory subjects in

Oskar Pfungst. *The first scientist to demonstrate cueing experimentally, Pfungst appears here several years after his work with Clever Hans.*

211

Berlin, and apparently it happened with Freud's private patients in Vienna.

A man of great dignity, imagination, and strength of conviction, Freud did not restrain himself in his early therapeutic sessions. He was convinced that every neurotic patient had undergone a seduction experience, and he awaited that remembrance. When he pressed physically on the patient's forehead, looking for a certain response, he pressed verbally for the same reply, a statement of childhood seduction. Not surprisingly, especially considering the trance-like conditions under which he worked, his patients often produced such "memories."

Little Hans' father, an early, enthusiastic supporter of psychoanalysis, was determined to help Freud in whatever way he could. Providing background information and questioning the boy at length, apparently he gave his son certain ideas consistent with psychoanalysis. At one point he even referred to himself as a white horse.

On another occasion he spoke to Little Hans about the horse that fell down, asking if it reminded Hans of his father.

"It's possible," the boy answered.

Then the father inquired about the black around the horse's mouth: "What? A mustache, perhaps?"

Hans replied, laughing, "Oh no!"

Freud felt the problem had been avoided by keeping the father unacquainted with his own ideas about the case. But the father certainly was aware of the sexual and aggressive themes in psychoanalytic thought, perhaps including the Oedipus hypothesis. And Freud himself once described the possibility of unconscious cueing. "No mortal can keep a secret," he said. "If his lips are silent, he chatters with his fingertips; betrayal oozes out of him at every pore."

The old schoolmaster in Berlin of course had a secret; he almost always knew the answer to the question asked of the horse. But Mr. von Osten could not keep that secret. His lips were silent, but he chattered with his upper body. Betrayal oozed out of him at every head movement, forward and backward.

And there is one final irony here. Sigmund Freud knew of Clever

Hans, just as Oskar Pfungst knew of his work. A friend had written to Freud of the wonderful Berlin horse that possessed a human consciousness, planning to visit the animal. Freud was immediately interested, urging his friend to spend two weeks studying Mr. von Osten's prodigy and all the available evidence. Then they would publish an account in Freud's journal.

Freud even proposed a title. The article would be something about the occult; it would be called "The Unconscious and Thought Transference." The redoubtable Sigmund Freud, at mid-career and fifty years of age, though he later changed his mind, apparently at that moment had forgotten all about the possibility of cueing and was instead considering the telepathy hypothesis in the case of Clever Hans.

Sigmund Freud. *Having described unconscious cueing in a clinical setting, Freud is shown here at the time he studied Little Hans.*

Our story of psychology from Berlin to Vienna has been dedicated to this theme of investigator effects. Pfungst and Freud as investigators, Clever and Little Hans as subjects, and Mr. von Osten and Little Hans' father as unwitting culprits all played a role in demonstrating various investigator effects. The old schoolmaster's subtle signals were discovered by the experimental method. The parent's subtle signals were observed in the clinical method. Both studies show the potential for errors not only through cueing but also in observation and interpretation. Control of these factors is a constant problem in science, and it has been recognized as a special problem in psychology since the early days of Clever and Little Hans.

In fact, while the horse was dutifully tapping for Mr. von Osten, laboratory dogs were uncontrollably salivating for their Russian master, Ivan Pavlov, and this outcome was also attributable to investigator effects. Pavlov had been interested in the digestive system and his dogs salivated at the sight of food, but as the trainer's footsteps sounded the arrival of the food each day, these noises themselves became sufficient to elicit salivation. Confronted with this phenomenon, Pavlov dropped his work in physiology and took up psychology, focusing on a topic now known as classical conditioning, concerned with the acquisition of learned reflexes and simple emotional reactions. The cueing in this case was unrelated to the investigator's expectations, for Pavlov was studying physiology, not mental processes, but the effect occurred nevertheless.

In recent years this problem has been reflected in humanity's long-standing desire to communicate with lower animals. Clever systems have been devised for teaching human languages to various species, especially chimpanzees, sometimes with impressive results. But the possibility of investigator effects remains. The foresighted Stumpf, eighty years ago, penned a warning in this regard, which even today merits close attention.

These diverse efforts by Pfungst and Freud and the others show that science is a human enterprise, a celebration of human achievement and a monument to human frailty. They emphasize that scientific inquiry is a function of the whole individual, influenced by the kind of person the scientist is, the expectations brought to the research, and the procedures employed for that purpose. The investigator's total being is involved, and the Hans legacy has been devoted to this purpose: to illustrating the inescapable human element in science.

The value of this lesson could hardly have been foreseen in psychology's early days, but since then it has proven useful not only in psychology but in other sciences as well, including those with a much longer history. Psychology in this way, through research on investigator effects and methods for dealing with them, has partially repaid its debt to its earliest ancestors, especially to biology, physics, and physiology, which contributed most significantly to its founding. Psychology first developed in the shadow of these fields, using some of their ideas and methods, and it has returned the favor through its study of the behavior of the scientist, as he or she goes about the process of human inquiry. The study of investigator effects, which began most dramatically with the Hans research at the turn of this century, is thus left as a legacy to the broad spectrum of all science, as repayment for earlier assistance.

Altogether, this story of psychology has illustrated unconscious signals and perhaps unconscious motivation, the systems of behaviorism and psychoanalysis, and research methods in experimental and clinical psychology. But most important, it has shown that psychology, through the study of human behavior, can contribute to a better understanding of the research process in all sciences. Therein lies the legacy.

The Horse and Boy

Epilogue

In the months and years following this research Clever Hans and Little Hans found their ways down different paths. One advanced along a road of public acclaim. The other pursued a more solitary journey. But which was which?

As things turned out, the boy called Hans, who finally mustered the courage to venture onto the street and into the world beyond, kept much to himself as an adult. Living in intentional obscurity in a city in Western Europe, he avoided all follow-up studies. Even if he were agreeable, any retrospective account would be of doubtful value, subject to the memory distortions of several decades.

He was seemingly unhindered by any further phobic condition, however. After the stormy earlier period, he apparently entered into a reasonably contented later life, having forgotten the whole previous episode. This report, though based upon superficial evidence, may give some further hope to parents and children in a similar condition.

The horse called Hans, which could not muster the intelligence to respond to his own name, was destined for another outcome. The public lost interest temporarily but not the horse's new owner, Mr. Karl Krall, the jeweler from Elberfeld, who knew that between the still-credulous and ever-curious there is always an audience for such performances. Under the tutelage of his affluent new proprietor, Clever Hans began to perform again for carrots and crusts of bread.

As the horse's success mounted, so did Krall's enthusiasm. This gentleman even attended a conference on learned animals at the University of Amsterdam and described Clever Hans' intelligence to that audience. He then bought two more horses, Muhamed and Zarif, and together with Clever Hans and some others, they kicked the proverbial dead horse into full life again. Within a few years they became known as the famous Horses of Elberfeld. Krall's demonstrations and his handsome new book about the intelligent Elberfeld horses were well received.

Muhamed was an expert at arithmetic. He learned simple numbers in a few days and was able to count within two weeks. In short order he mastered multiplication, division, and fractions, and then became expert in extracting square and cube roots. Zarif's talent lay in spelling, which he often accomplished with special variations.

Perhaps Mr. Krall's success lay in his classroom methods, for numbers were tapped more efficiently, the right foot for hundreds, the left foot for tens, and the right foot again for units. A better spelling tablet was prepared, and the horses had their own inclined desk on which to tap. And to maintain proper discipline the rod was not spared.

Nevertheless, there was room for doubt. The new procedures raised some further questions.

The horse performed in darkness, showing that they were not receiving signals—but of course there was enough light to make the scene interesting for spectators. The horses answered difficult questions as readily as easier ones—but Mr. Krall said that was because they were so intelligent. The horses sometimes performed poorly—but these tests were made by outsiders.

When the horses failed, Mr. Krall explained that they were

shedding their coats. When excellent results were obtained a few days later, he announced that the questioner must have the horses' confidence. Indeed, the horses sometimes performed so well that they did not need to look at the numbers and letters on the chart.

So Clever Hans with the other Elberfeld horses, went on tapping. In recognition of this resurrection, one journalist wrote:

"You, Clever Hans, who earned a nation's praise,
Seemed fated, sadly, so to end your days
That pots of glue your spirit would contain.
But now your latest coup *shows quite a brain.*
You're pensioned off; go graze, live long, get fat.
Not every ass can pull a stunt like that."

It was said that mischievous grooms were responsible, for one stableboy was observed to open and close his eyes when an Elberfeld horse began and ceased tapping. Another was noted to alter his hold on the halter when answers were required. On still another occasion, as the horse said "no" by moving his head from side to side, the groom seemingly pinched the animal first on the right and then on the left flank. The jeweler may have been a man of his word, but his assistants perhaps were not so trustworthy. In any case, their roles in the performance of the Elberfeld horses were never fully put to a scientific test.

References

Each source appears in order of use in this text, as indicated by the numbers preceding the entry. Thus, 10:22 refers to page 10, line 22, and the relevant discussions begin at that point.

Unless otherwise cited, the source for Clever Hans is Pfungst, *The Horse of Mr. von Osten.* For Little Hans it is Freud, "Analysis of a Phobia" in the *Standard Edition of the Complete Works,* Volume 10.

Foreign works generally appear in the original language. The chief translator for publications in German has been Professor Thomas Hansen, German Department, Wellesley College, Wellesley, Massachusetts.

CHAPTER 1

7:1   Sanford, E. C. Psychic research in the animal field: Der kluge Hans and the Elberfeld horses. *American Journal of Psychology,* 1914, *25,* 1–31.

7:7   Pfungst, O. *The Horse of Mr. von Osten (Clever Hans).* Trans. by Rahn, C. L. New York: Holt, 1911. See also Rosenthal, R. (Ed.) *Clever Hans: The Horse of Mr. von Osten.* Trans. by Rahn, C. L. New York: Holt, 1965.

8:14 Equine prodigy knows music and arithmetic. *New York Times,* August 14, 1904, p. 4.

8:28 Freud, S. Analysis of a phobia in a five-year-old boy. In *Standard Edition of the Complete Works of Sigmund Freud,* Vol. 10. London: Hogarth Press, 1953.

10:3 Darwin, C. *The Origin of the Species by Means of Natural Selection* (6th ed.). New York: Appleton, 1972.

10:22 Boring, E. G. *A History of Experimental Psychology* (2nd ed.). New York: Appleton-Century-Crofts, 1950. See also Stumpf, C. Carl Stumpf. In Murchison, E. (Ed.) *A History of Psychology in Autobiography.* 1. Worcester, Massachusetts: Clark University Press, 1930.

12:29 Angell, J. R. Prefatory note. In Pfungst, O. *The Horse of Mr. von Osten* (*Clever Hans*). Trans. by Rahn, C. L. New York: Holt, 1911.

12:34 Freud, S. Analysis of a phobia.

CHAPTER 2

17:1 Becker, C. L. & Cooper, K. S. *Modern History.* Morristown, New Jersey: Silver Burdett, 1977.

19:17 Pfungst, O. *Horse of Mr. von Osten.*

21:26 Fernald, L. D. & Fernald, P. S. *Introduction to Psychology* (4th ed.). Boston: Houghton Mifflin, 1978.

22:19 Boring, E. G. *History of Experimental Psychology.*

22:37 Pfungst, O. *Horse of Mr. von Osten.*

23:7 Krall, K. *Denkende Tiere.* Leipzig: Verlag von Friedrich Engelmann, 1912.

23:10 Hans, the wonderful Orloff stallion. *Country Calendar, 1,* May, 1905, p. 90.

23:15 Heyn, E. T. Berlin's wonderful horse. *New York Times,* September 4, 1904, p. 17.

24:1 Krall, K. *Denkende Tiere.*

24:14 Stumpf, C. Introduction. In Pfungst, O. *The Horse of Mr. von Osten* (*Clever Hans*). Trans. by Rahn, C. L. New York: Holt, 1911.

24:27 Heyn, E. T. Berlin's wonderful horse.

24:36 Hans, the wonderful Orloff stallion.

25:1 Krall, K. *Denkende Tiere.*

26:12   Block, P. Der kluge Hans. *Berliner Tageblatt,* August 15, 1904, p. 1.

26:16   Equine prodigy knows music and arithmetic. New York Times.

26:26   Heyn, E. T. Berlin's wonderful horse.

27:18   Boring, E. G. *History of Experimental Psychology.*

28:22   Krall, K. *Denkende Tiere.*

28:35   Freund, F. (Amicus) *Der "kluge" Hans? Ein Bertrag zur Aufklärung.* Berlin, 1904. See Krall, K. *Denkende Tiere.*

28:36   Zell, T. *Das rechnende Pferd. Ein Gutachten über den "Klugen Hans" auf Grund eigener Beobachtungen.* Berlin, 1904. See Krall, K. *Denkende Tiere.*

30:1   Krall, K. *Denkende Tiere.*

30:32   Stumpf, C. Supplements. In Pfungst, O. *The Horse of Mr. von Osten (Clever Hans).* Trans. by Rahn, C. L. New York: Holt, 1911.

31:5   Boring, E. G. *History of Experimental Psychology.*

32:25   Pfungst, O. *Horse of Mr. von Osten.* See also Stumpf, C. Supplements.

34:6   Stumpf, C. Supplements.

CHAPTER 3

35:6   Stumpf, C. Introduction. See also Stumpf, C. Supplements.

36:30   Herrnstein, R. J. Introduction. In Watson, J. B. *Behavior: An Introduction to Comparative Psychology.* New York: Holt, 1967. See also Skinner, B. F. *About Behaviorism.* New York: Knopf, 1974.

37:4   Stumpf, C. Supplements.

38:8   Rachlin, H. *Introduction to Modern Behaviorism.* San Francisco: W. H. Freeman, 1976.

38:24   Heyn, E. T. Berlin's wonderful horse.

38:33   Stumpf, C. Supplements.

CHAPTER 4

47:1   Stumpf, C. Introduction. See also Stumpf, C. Supplements.

48:3   Blondlot, R. L' existence de radiation solaire capable de traverser les metaux, le bois, etc. *Journal de Physique,* 1903, *2, 2s.*

See also Blondlot, R. Rayons X et rayons N. *Archchives des Science Physiques et Naturales*, 1904, *17*, 4s.

49:11 Charpentier, P. M. Phenomenes divers de transmission de rayons N et applications. *Academie Comptes Rendus*, 1904, *138*, 3s. See also Rostand, J. *Error and Deception in Science: Essays on Biological Aspects of Life*. Trans. by Pomerans, A. J. New York: Basic Books, 1960.

51:1 Stumpf, C. Introduction.

51:12 Equine prodigy knows music and arithmetic. *New York Times*.

51:16 Block, P. Der kluge Hans.

52:3 Swift, J. *Gulliver's Travels*. New York: Norton, 1961.

52:32 Sidgwick, H. Presidential address. *Proceedings of the Society for Psychical Research*, 1908, *22*, 57.

53:26 Pfungst, O. *Horse of Mr. von Osten*.

54:17 Hansel, C. E. *ESP: A Scientific Evaluation*. New York: Scribner, 1966.

55:15 Munsterberg, H. My friends, the spiritualists. *Metropolitan Magazine*, 1910, *31*, 558–572.

57:33 Krall, K. *Denkende Tiere*.

58:8 Stumpf, C. Introduction.

CHAPTER 5

61:8 Rostand, J. *Error and Deception in Science: Essays on Biological Aspects of Life*. Trans. by Pomerans, A. J. New York: Basic Books, 1960.

62:5 Wood, R. W. The N-Rays. *Nature*, 1904 *70*, 530–531.

64:2 Fernald, L. D. & Fernald, P. S. *Introduction to Psychology*.

65:5 Pfungst, O. *Horse of Mr. von Osten*.

68:15 Fernald, L. D. & Fernald, P. S. *Introduction to Psychology*.

69:12 Pfungst, O. *Horse of Mr. von Osten*.

71:11 Stumpf, C. Der "Kluge Hans" und sein Professor. *Berliner Tageblatt*, December 10, 1904, p. 4.

71:14 Notes. *Nature*, 1904, *70*, 510.

71:17 *Berliner Morgenpost*, August 31, 1904.

71:21 Meehan, J. The Berlin "Thinking" horse. *Nature*, 1904, *70*, 602–603.

71:29   Stumpf, C. Introduction.

72:1   Pfungst, O. *Horse of Mr. von Osten.*

75:3   Hans does not think but reads expression. New York Times, December 11, 1904, p. 4.

75:29   Krall, K. *Denkende Tiere.*

CHAPTER 6

78:11   Pfungst, O. *Horse of Mr. von Osten.* See also Stumpf, C. Introduction.

84:13   Simpson, G. G. *Horses: The Story of the Horse Family in the Modern World.* New York: Oxford University Press, 1951.

85:12   Pfungst, O. *Horse of Mr. von Osten.*

88:10   Dessoir, M. A miracle? *Berliner Tageblatt,* August 29, 1904.

88:34   Lubbock, J. *On the Senses, Instincts and Intelligence of Animals.* London: Kegan Paul, 1888.

89:4   Kretschmer, H. Letter. *Schlesische Zeitung,* Breslau, August 21, 1904. See Pfungst, O. *Horse of Mr. von Osten.*

89:15   Tolstoy, L. *Anna Karenina.* Trans. by Edmonds, R. New York: Penguin, 1978.

89:26   Pfungst, O. *Horse of Mr. von Osten.*

CHAPTER 7

95:3   Stumpf, C. Supplements.

97:24   Mees, C. E. Scientific thought and social reconstruction. *Electrical Engineering,* 1934, 53, 383–384.

99:10   Polanyi, M. *Personal Knowledge: Towards a Post-critical Philosophy.* University of Chicago Press, 1958. See also Popper, K. R. *The Logic of Scientific Discovery* (2nd ed.). New York: Harper, 1968.

99:27   Jones, E. *The Life and Work of Sigmund Freud.* Vol. 1. New York: Basic Books, 1953.

CHAPTER 8

103:9   Janik, A. & Toulmin, D. *Wittgenstein's Vienna.* New York: Simon & Schuster, 1973. See also Schorske, C. E. *Fin-de-Siecle Vienna: Politics and Culture.* New York: Vintage Books, 1981.

105:13   Freud, S. Analysis of a phobia.

107:7   Fernald, L. D. & Fernald, P. S. *Introduction to Psychology.*

109:9   Freud, S. Analysis of a phobia.

110:22   Starr, M. A. Memorial to Professor Jean-Marie Charcot. *International Clinics*, 1894, *1*, ix–xxi.

111:18   Freud, S. Analysis of a phobia.

CHAPTER 9

117:1   Freud, S. Analysis of a phobia.

118:30   Shirley, M. M. The first two years. *Child Welfare Monograph*, No. 7. Minneapolis: University of Minnesota Press, 1933.

120:31   Halverson, H. M. The development of prehension in infants. In Barker, R. D. *et al.*, *Child Behavior and Development*. New York: McGraw-Hill, 1943.

121:34   Freud, S. Analysis of a phobia.

125:13   Fernald, L. D. & Fernald, P. S. *Introduction to Psychology.*

126:4   Freud, S. Analysis of a phobia.

126:22   Piaget, J. *Six Psychological Studies*. New York: Random House, 1967. See also Piaget, J. & Inhelder, B. *Psychology of the Child*. London: Routledge & Kegan Paul, 1969.

126:35   Freud, S. Analysis of a phobia.

CHAPTER 10

131:1   Jones, E. *Life and Work of Sigmund Freud*, Vol. *1*.

138:27   Freud, S. Three essays on the theory of sexuality. In *Standard Edition of the Complete Works of Sigmund Freud*, Vol. *12*. London: Hogarth, 1953.

140:20   Freud, S. Repression. In *Standard Edition of the Complete Works of Sigmund Freud*, Vol. *14*. London: Hogarth, 1953.

140:31   Gay, P. Introduction. In Englemann, E. *Berggasse 19: Sigmund Freud's Home and Offices in Vienna, 1938*. New York: Basic Books, 1976.

142:19   Freud, S. Three essays on the theory of sexuality.

143:12   Freud, S. The ego and the id. In *Standard Edition of the Complete Works of Sigmund Freud*, Vol. *19*. London: Hogarth, 1953.

144:17   Jones, E. *Life and Work of Sigmund Freud*, Vol. *1*.

144:23   Grinstein, A. *On Sigmund Freud's Dreams*. Detroit: Wayne University Press, 1968.

CHAPTER 11

147:1   Freud, S. Analysis of a phobia.

148:31   Jung, C. G. The association method. *American Journal of Psychology*, 1910, *21*, 219–269.

150:1   Freud, S. Analysis of a phobia.

151:33   Freud, S. The interpretation of dreams. In *Standard Edition of the Complete Works of Sigmund Freud*, Vols. 4, 5. London: Hogarth, 1953.

156:1   Freud, S. Analysis of a phobia.

159:16   Axline, V. *Play Therapy* (rev ed.). New York: Ballantine, 1974.

159:29   Freud, S. Analysis of a phobia.

CHAPTER 12

163:6   Freud, S. Analysis of a phobia.

165:1   Jones, E. *The Life and Work of Sigmund Freud*, Vol. 2. New York: Basic Books, 1955.

165:8   Jung, C. G. The association method.

168:5   Jung, C. G. *Memories, Dreams, Reflections*. New York: Pantheon Books, 1963.

168:20   Jung, C. G. *Contributions to Analytical Psychology*. New York: Harcourt, 1928.

169:11   Adler, A. *Practice and Theory of Individual Psychology*. Trans. by Rodin, P. Atlantic Highlands, New Jersey: Humanities, 1971.

170:19   Garrison, M. A new look at Little Hans. *Psychoanalytic Review*, 1978, *65*, 523–532.

170:19   Maurer, A. Did Little Hans really want to marry his mother? *Journal of Humanistic Psychology*, 1964, Fall, 139–148.

170:20   Wile, J. R. Freud's treatment of Little Hans: The case of an undeveloped family therapist. *Family Therapy*, 1980, *7*, 131–138.

170:22   Robbins, W. S. Varieties of Oedipal distortions in severe character pathologies. *Journal of American Psychoanalytic Asso-*

ciation, 1977, 25, 201–218. See also Loewald, H. W. The waning of the Oedipus complex. *Journal of American Psychiatric Association*, 1979, 27, 751–775.

171:5   Malinowski, B. *Sex and Repression in Savage Society.* New York: Harcourt Brace, 1927. See also Malinowski, B. *The Sexual Life of Savages: In North-Western Melanesia.* New York: Harcourt Brace, 1929.

173:6   Eggan, D. The General Problem of Hopi Adjustment. In Kluckholm, C. and Murray, H. A. (Eds.) *Personality in Nature, Society and Culture.* New York: Knopf, 1953.

174:6   Valabrega, J. P. L'anthropologie psychanalytique, *Psychanalyse,* 1957, 3, 221–245.

174:6   Roheim, G. The Anthropological Evidence and the Oedipus Complex, *Psychoanalytic Quarterly,* 1952, 21, 537–542.

174:7   Devereaux, G. The Oedipal Situation and Its Consequences in the Epics of Ancient India. *Samiksa,* 1951, 5, 5–13.

174:7   Papertian, G. The Rubicon Complex: Incest in Ancient Civilizations, *Confinia Psychiatrica,* 1972, 15, 116–124.

174:10   Stephens, W. N. *The Oedipus Complex: Cross-cultural Evidence.* Glencoe, IL: The Free Press, 1962.

174:24   Miller, A. R. Analysis of the Oedipal Complex, *Psychological Reports,* 1969, 24, 781–782.

175:17   Freidman, S. M. An Empirical Study of the Castration and Oedipus Complexes, *Genetic Psychology Monographs,* 1952, 46, 61–130.

176:16   Wolpe, J. & Rachman, S. Psychoanalytic "Evidence:" A critique based on Freud's case of Little Hans. *Journal of Nervous and Mental Disease,* 1960, 130, 135–148.

177:4   Freud, S. Analysis of a phobia.

177:33   Jones, E. *Life and Work of Sigmund Freud,* Vol. 1.

178:7   Freud, S. Analysis of a phobia.

CHAPTER 13

184:1   Pfungst, O. *Horse of Mr. von Osten.*

186:36   Krall, K. *Denkende Tiere.*

188:13   Pfungst, O. *Horse of Mr. von Osten.*

189:10   Freud, S. Analysis of a phobia.

190:1   Cromer, W. & Anderson, P. Freud's visit to America: Newspaper coverage. *Journal of the History of Behavioral Science*, 1970, 6, 349–353.

190:12   Hale, N. G. *Freud and the Americans: The Beginnings of Psychoanalysis in the United States, 1876–1917*. New York: Oxford University Press, 1971.

190:19   Freud, S. Analysis of a phobia.

191:15   American Psychiatric Association Task Force on Nomenclature and Statistics. *Diagnostic and Statistical Manual: Mental Disorders III*. Washington, D.C.: American Psychiatric Association, 1980.

192:8   Jones, E. *Life and Work of Sigmund Freud*, Vol. 2.

CHAPTER 14

195:1   Pfungst, O. *Horse of Mr. von Osten*.

196:31   Krall, K. *Denkende Tiere*.

197:7   Pfungst, O. *Horse of Mr. von Osten*.

198:12   Scheflen, A. E. The significance of posture in communication systems, *Psychiatry*, 1964, 27, 316–331. See also: Timaeus, E. *Experiment und Psychologie*. Göttingen: Verlag für Psychologie Hogrefe, 1974; Ekman, P. & Friesen, W. V. Nonverbal leakage and clues to deception, *Psychiatry*, 1969, 32, 88–105; Scheflen, A. E. Communication and regulation and psychotherapy, *Psychiatry*, 1963, 26, 126–136.

198:23   Krall, K. *Denkende Tiere*.

199:3   Jones, E. *The Life and Work of Sigmund Freud*, Vol. 3. New York: Basic Books, 1957.

201:11   Johnson, H. M. The talking dog, *Science*, 1912, 35, 749–751.

202:2   Beveridge, W. I. B. *The Art of Scientific Investigation*. New York: Random House, 1957.

203:3   Beveridge, W. I. B. *The Art of Scientific Investigation*.

203:4   Jones, E. *Life and Work of Sigmund Freud*, Vol. 1.

205:24   Boring, E. G. Eponym as placebo. In *History, Psychology and Science: Selected Papers*. New York: Wiley, 1963.

206:26   Rosenthal, R. Introduction. Clever Hans: A case study of scientific method. In Rosenthal, R. (Ed.) *Clever Hans: The Horse of Mr. von Osten*. Trans. by Rahn, C. L. New York: Holt, 1965.

CHAPTER 15

207:12 Rosenthal, R. *Experimenter Effects in Behavioral Research.* New York: Halstead Press, 1976.

208:1 Pfungst, O. *Horse of Mr. von Osten.* See also Stumpf, C. Introduction.

208:24 Blondlot, R. Rayons X et rayons N. *Archives des Sciences Physiques et Naturales,* 1904, *15,* 4s. See also Rostand, J. *Error and Deception in Science.*

209:9 Rosenthal, R. *Experimenter Effects in Behavioral Research.*

209:18 Pfungst, O. *Horse of Mr. von Osten.*

209:32 Krall, K. *Denkende Tiere.* See also Stumpf, C. Introduction.

210:6 Jones, E. *Life and Work of Sigmund Freud,* Vol. *1.*

210:19 Gelman, D. Finding the hidden Freud. *Newsweek,* November 30, 1981, 64–70. See also Krüll, M. *Freud and His Father.* In press.

210:28 Leo, J. Muffling the master's voice. *Time,* November 23, 1981, p. 59.

210:35 Rosenthal, R. *Experimenter Effects in Behavioral Research.*

212:18 Freud, S. Analysis of a phobia.

212:29 Freud, S. Fragment of an analysis of a case of hysteria. In *Standard Edition of the Complete Works of Sigmund Freud,* Vol. *7,* London: Hogarth, 1953.

213:1 Jones, E. *Life and Work of Sigmund Freud,* Vol. *3.*

214:13 Pavlov, I. P. *Conditional Reflexes.* Trans. by Anrep, G. V. New York: Oxford University Press, 1927. See Gruenberg, B. *The Story of Evolution.* Princeton, New Jersey: Van Nostrand, 1929.

214:26 Stumpf, C. Supplements. See also Sebeok, T. A. & Umiker-Sebeok, J. Performing animals: Secrets of the trade. *Psychology Today,* November, 1979, 78–91; Sebeok, T. A. & Umiker-Sebeok, J. (Eds.) *Speaking of Apes.* New York: Plenum Press, 1980.

EPILOGUE

217:5 Freud, A. Forward. In Gardiner, M. *The Wolf Man.* New York: Basic Books, 1971.

218:3 Sanford, E. C. Psychic research in the animal field.

218:9   Jordan, R. *Der "Entlarver des klugen Hans" — Entlarvt!*
          Grother: Deutscher Tierfreund, 1932.

218:12   Sanford, E. C. Psychic research in the animal field.

219:7   *Lustige Blatter.* Nv. 30, 1909. In Krall, K. *Denkende Tiere.*

219:13   Sanford, E. C. Psychic research in the animal field.

Illustrations

All illustrations, except for the composite artwork facing the epilogue, are from original photographs, maps, or diagrams. Each location in this text is indicated by the page number in parentheses, which follows the title and precedes the source.

City Walkway (2) Wechsberg, J. *Vienna, My Vienna.* New York: Macmillan, 1968.

Griebenow Courtyard (4) Krall, K. *Denkende Tiere.* Leipzig: Verlag von Friedrich Engelmann, 1912.

Wundts Upon a Birthday (11) Fernald, L. D. & Fernald, P. S. *Introduction to Psychology,* 4th edition. Boston: Houghton Mifflin, 1978.

Near Unter den Linden (14) Titzenthaler, W. *Berlin: Photographen des 19 Jahrhunderts.* Berlin: Rembrandt Verlag, 1968.

Changing Times (18) Titzenthaler, W. *Berlin: Photographen des 19 Jahrhunderts.*

Crowds in the Courtyard (25) Krall, K. *Denkende Tiere.*

Front Page News (29) *Berliner Tageblatt,* August 29, 1904.

Carl Stumpf (31) *American Journal of Psychology,* 49, 1937.

Teacher and Pupil (37) Pfungst, O. *The Horse of Mr. von Osten (Clever Hans)*. Trans. by Rahn, C. L. New York: Holt, 1911.

Tapping Response (40) Krall, K. *Denkende Tiere.*

Work Tables (43) Krall, K. *Denkende Tiere.*

Institute at Nancy (48) Postcard Collection. Fogg Art Museum, Harvard University, Cambridge, Massachusetts.

Charles Darwin (50) Moorehead, A. *Darwin and the Beagle.* New York: Harper & Row, 1969.

The Journal (51) Moorehead, A. *Darwin and the Beagle.*

Madam Palladino (56) Munsterberg, H. My friends, the spiritualists. *Metropolitan Magazine*, 1910, *31*, 558–572.

Courtyard Neighbors (63) Krall, K. *Denkende Tiere.*

The Blinder (67) Krall, K. *Denkende Tiere.*

Testing Clever Hans (73) Krall, K. *Denkende Tiere.*

Recording Apparatus (82) Fernald, L. D. & Fernald, P. S. *Introduction to Psychology*, 4th edition. Boston: Houghton Mifflin, 1978.

The Signals (83) Pfungst, O. *Horse of Mr. von Osten.*

Visual Fields (84) Walls, G. L. Eye movements and the fovea. In Walls, G. L. *The Vertebrate Eye and Its Adaptive Radiation.* Bloomfield Hills, Michigan: Cranbrook Institute of Science, 1942.

Hans' Eyes (86) Krall, K. *Denkende Tiere.*

Military Parade (90) Titzenthaler. W. *Berlin: Photographen des 19 Jahrhunderts.*

The Horse's Mouth (92) Henry, M. *All About Horses.* New York: Random House, 1962.

Horses Teeth (98) Krall, K. *Denkende Tiere.*

Ringstrassen Cafe (100) Weigel, H. *Das Wiener Kaffeehaus.* Wien: Verlag Fritz Molden, 1978.

City Landmark (104) Gerlach, M. *Wien: Eine Auswahl von Stadtbildern.* Herausgegeben von der Gemeinde Wien, 1918.

Schotten Ring Traffic (106) Photograph Collection. Fogg Art Museum, Harvard University, Cambridge, Massachusetts.

Horses in Harness (114) Gorny, H. *Ein Pferdbuch.* München: F. Bruckmann, 1938.

Early Childhood (119) Personal Collection. Professor Harry Zohn, Department of Germanic and Slavic Language, Brandeis University, Waltham, Massachusetts.

Setting for a Dream (129) Photograph Collection, Architecture. Fogg Art Museum, Harvard University, Cambridge, Massachusetts.

19 Berggasse (133) Freud, E., Freud, L. & Grubrich-Simitis, I. *Sigmund Freud.* New York: Harcourt Brace Jovanovich, 1976.

The Couch (135) Englemann, E. *Berggasse 19 — Sigmund Freud's Home and Offices in Vienna, 1938.* New York: Basic Books, 1976.

Freud's Relics (141) Freud, E., Freud, L. & Grubrich-Simitis, I. *Sigmund Freud.*

Father and Son (145) Cohen, J. *Personality Dynamics.* Chicago: Rand McNally, 1969.

Freud's Desk (149) Freud, E., Freud, L. & Grubrich-Simitis, I. *Sigmund Freud.*

At the Zoo (155) Peterman, R. E. *Wien im Zeitalter Kaiser Franz Josephs I.* Wien: Verlag R. Lechner, 1908.

Haupzollamt Station (158) Photograph Collection, Architecture. Fogg Art Museum, Harvard University, Cambridge, Massachusetts.

On the Promenade (160) Gerlach, M. *Wien: Eine Auswahl von Stadtbildern.*

Carl Gustav Jung (164) Cohen, J. *Personality Dynamics.*

The Adlers (169) Rattner, J. *Alfred Adler in Selbstzeugnissen und Bildokumenten.* Hamburg: Rowohlt, 1972.

Melanesians at Play (171) Malinowski, B. *Sex and Repression in a Savage Society.* New York: Harcourt Brace, 1927.

Bachelor's Hut (173) Malinowski, B. *Sex and Repression in a Savage Society.*

Cafe Conversations (178) Weigel, H. *Das Wiener Kaffeehaus.*

19 Berggasse (180) Frontispiece. *The Standard Edition of the Complete Works of Sigmund Freud,* Vol. 7. London: Hogarth, 1953.

Color Test (185) Krall, K. *Denkende Tiere.*

Diagram of the Griebenow Courtyard (187) Krall, K. *Denkende Tiere.*

Der Kluge Hans (188) Pfungst, O. *Horse of Mr. von Osten.*

Der Kleine Hans (189) *Jahrbuch fur Psychoanalytische und Psychopathologische Forschungen,* Vol. 1, 1909.

Diagram of Lower Viaduct Street (191) Freud, S. Analysis of a phobia in a five-year-old boy. In *The Standard Edition of the Complete Works of Sigmund Freud,* Vol. 10. London: Hogarth, 1953.

Freud in the Wilds (192) Gifford, G. E. Freud and the porcupine. *Harvard Medical Alumni Bulletin,* 1972, 46, 28–31.

Wilhelm von Osten (197) Krall, K. *Denkende Tiere.*

Anna O. (199) Freeman, L. The immortal Anna O.—From Freud to feminism. *New York Times Magazine,* November 11, 1979, p. 30.

Pfungst and the Horse (202) Krall, K. *Denkende Tiere.*

Sigmund and Martha (203) Freud, E., Freud, L. & Grubrich-Simitis, I. *Sigmund Freud.*

Oskar Pfungst (211) Sebeok, T. A. & Umiker-Sebeok, J. (Eds.) *Speaking of Apes.* New York: Plenum Press, 1980.

Sigmund Freud (213) Freud, E., Freud, L. & Grubrich-Simitis, I. *Sigmund Freud.*

The Horse and Boy (216) Composite illustration from Krall, K. *Denkende Tiere* and from the personal collection of Professor Harry Zohn, Department of Germanic and Slavic Language, Brandeis University, Waltham, Massachusetts.

INDEX

Traditional psychoanalysis. *See*
  Psychoanalysis
Triskaidekophobia, 108

U

Unconscious motivation, 142
Unverified hypothesis, 176

V

Variables
  dependent, 68
  independent, 68
Verbal items, 27
Verification, 78
Visual field, 84
Vision
  binocular, 84
  monocular, 84
  size of field, 84

von Osten, Wilhelm, 24–46, 49, 51,
  53, 59–76, 78, 185, 196–198,
  209, 212

W

Wallace, Alfred R., 206
Wood, R. W., 62
Word association test, 148
Wundt, Wilhelm, 11, 32

X

X rays, 48

Y

Youth, Melanesian versus European,
  171–173

Z

Zoophobia, 108